T0337955

Sand Production Management for Unconsolidated Sandstone Reservoirs

Sand Production Management for Unconsolidated Sandstone Reservoirs

Shouwei Zhou

Academician of Chinese Academy of Engineering, Former Vice President of China National Offshore Oil Corporation (CNOOC), China

Fujie Sun

General Manager of Technology Development Department of CNOOC, China

WILEY Blackwell

This edition first published 2016

Published by John Wiley & Sons Singapore Pte. Ltd., 1 Fusionopolis Walk, #07-01 Solaris South Tower, Singapore 138628, under exclusive license granted by Petroleum Industry Press for all media and languages excluding Simplified and Traditional Chinese and throughout the world excluding Mainland China, and with non-exclusive license for electronic versions in Mainland China.

For details of our global editorial offices, for customer services and for information about how to apply for permission to reuse the copyright material in this book please see our website at www.wiley.com.

Library of Congress Cataloging-in-Publication Data

Names: Zhou, Shouwei. | Sun, Fujie.
Title: Sand production management for unconsolidated sandstone reservoirs / Shouwei Zhou and Fujie Sun.
Description: Singapore : John Wiley & Sons, Inc., 2016. | Includes bibliographical references and index.
Identifiers: LCCN 2015034588 | ISBN 9781118961896 (cloth)
Subjects: LCSH: Oil sands. | Petroleum–Prospecting. | Petroleum–Geology.
Classification: LCC TN870.54.Z46 2016 | DDC 622/.3383–dc23
LC record available at http://lccn.loc.gov/2015034588

Cover credit: Off shore oil platforms at sunset © Getty Images
Printed and bound in Malaysia by Vivar Printing Sdn Bhd

1 2016

Contents

Foreword

Poorly consolidated heavy oil sandstone reservoirs represent a large and increasing percentage of China's oil and gas resources. The reasonable and efficient development of these reservoirs is becoming one of the important growth points of China's oil industry. The economic development of poorly consolidated reservoirs will significantly maintain and improve China's oil production rate and reduce production cost.

The features of unconsolidated heavy oil sandstone reservoirs are shallow depth, poor cementation, and high oil viscosity. One major trouble associated with these unconsolidated heavy oil sandstone pays is that particles in the formation are prone to move and thus sand production will occur. The occurrence of sand production will bring damage to the well and facilities and result in economic loss; in addition, the regular production of the oilfield will be affected. Although the approach of "complete sand control" can reduce the rate of sand production, the oil production rate is also affected and is reduced severely. As a result, the philosophy of sand production management was brought forward. Essentially, sand production management aims to extend the maximum, allowable, and safe sand production rate. The approach of sand production management is to optimize the production rate when considering both situations of "complete sand control" and production with limited sand production. In other words, the decision-making and production strategies are based on the consideration of well productivity index, wellbore sand-carrying capability, surface handling facilities, and so on between the conditions of complete sand control and limited sand production.

This book presents a comprehensive introduction of technologies applied in the development of poorly consolidated sandstone reservoirs. The contents include studies of sand production mechanisms, sand production prediction and productivity index appraisal, selection of well completion approaches, artificial lift technologies, and the implementation workflow of sand production management. The new ideas, philosophy, creative technologies, and operating and managing models are summarized. Technologies and competence to develop offshore unconsolidated heavy oil sandstone reservoirs in China are presented.

The achievements in this book are the collections of studies and applications of researchers in the domain of offshore oilfield exploitation for many years. The contents provide a new approach to develop unconsolidated heavy oil reservoirs. Publishing this information will be beneficial to other researchers, engineers, and on-site operation management staff who are involved in oil exploitation.

Luo Pingya
Academician of Academy of Engineering Chinese;
Former President of Southwest Petroleum University

Preface

Based on the on-site production requirements, sand production management is a new approach that is somewhere between cold heavy oil production and sand control. Compared with the conventional sand control approach, sand production management can not only reduce well completion cost but also increase well production rate by encouraging limited sand production of the pay. Contents of sand production management cover sand production prediction, well productivity appraisal, well completion, and wellbore sand-carrying aspects. The proposal of sand production management will be meaningful for the optimization of well completion design, rational well operating parameters and oilfield development, extension of economic development cycle of oil and gas wells, and other areas.

With the practical point of view, preliminary research on the potential problems and the corresponding solutions are conducted when developing unconsolidated sandstone reservoirs on the basis of theoretical studies and practices, the combination of reservoir engineering, and well completion engineering. Also, the ideas and approaches of how to improve the development of these reservoirs and economical results are proposed, which are useful for managers and engineers as reference and helpful for the cost-effective development of unconsolidated sandstone reservoirs.

Six chapters are included. Chapter 1 introduces the geological features of the formations of unconsolidated sandstone reservoirs and the concept of sand production management. Chapter 2 presents sand production mechanisms and the changes in formation properties due to sand production. Chapter 3 introduces the special approaches of estimating sand production rate and the solid–fluid coupling model of predicting sand production. Chapter 4 states the well completion technologies suited for sand production management, the approaches to sand control, and the designing methods of sand control parameters. Chapter 5 includes the critical flowing velocity of carrying sands, artificial lift technologies, and surface treatment of crude mixed with produced sands when producing with limited sands. Chapter 6 summarizes the applications of sand production management in Bohai Bay and presents the workflow involved in the execution of sand production management in LD 5-2 Oilfield.

Sand production management is a complex and integrated approach that involves many technologies, so the main research focus is on reservoir engineering, well completion technologies, and artificial lift. The related supporting technologies that will be used in sanding management are also briefly introduced.

Many research studies on sand production management were conducted for the development of unconsolidated sandstone reservoirs in recent years, which were supported by China National Offshore Oil Corporation (CNOOC), CNOOC (China) Ltd., and the Research Center of CNOOC. Special thanks are extended to the related departments and colleagues for all your support and assistance.

Sand production management is a new technology with a wide scope of knowledge and technology, and it is a subject that is continually evolving. We have tried our best to update, enrich, and improve the theoretical studies and practices, but this book is not free of errors or omissions. As such, your comments and insights are welcome and important for future improvement.

Shouwei Zhou
Fujie Sun

1

Introduction

1.1 Geological Characteristics of Unconsolidated Heavy Oil Sandstone Reservoirs

1.1.1 Distribution Characteristics of Unconsolidated Heavy Oil Sandstone Reservoirs

In 1992, Penbeyrt and Shaughnessy pointed out that 70 percent of the world's oil and gas resources are present in poorly consolidated sandstone reservoirs. These unconsolidated pays are distributed in nearly every oilfield around the world, such as Wilmington Oilfield and Kern River Oilfield in California, Bell Creek Oilfield in Montana, and S.E. Pauls Valley Oilfield in Oklahoma. Oilfields such as Cold Lake, Elk Point, Lindbergh, Lloylminster, Frog Lake, and Cactus Lake in Canada and some Paleogene-Neogene reservoirs in Indonesia, Trinidad, and Venezuela belong to unconsolidated reservoirs. In China, unconsolidated reservoirs are distributed in almost all existing oilfield areas (except Sichuan Basin), such as Shengli Oilfield, Dagang Oilfield, Bohai Oilfield in Bohai Bay, Liaohe Oilfield in Songliao Basin, and Karamayi Oilfield in Junggar Basin.

One feature of these reservoirs is a shallow depth, usually less than 1800 m. The shallowest ones are outcropping or buried under surface of tens of meters or 100 m of Paleogene–Neogene pay. Outside China, unconsolidated reservoirs also exist, such as heavy oilfields in Alberta, Canada. From the view of the structure, poorly consolidated heavy oil reservoirs are mainly distributed in the shallow pay of the slight dipping slope of basin or depression, the edge of salient, upper part

Sand Production Management for Unconsolidated Sandstone Reservoirs, First Edition.
Shouwei Zhou and Fujie Sun.

of flat salient, or faulted anticlinal trap of depression. These reservoirs have rich faults and are dominated by anticlinal structure reservoirs or structural nose reservoirs (Table 1-1). Edge water or bottom water is formed in these unconsolidated reservoir (especially in oilfields outside China), and some of them have a gas cap. Due to the features of shallow depth and poor cementation, porosity of these pays is commonly greater than 25%. Most of the heavy oil reserves are in this kind of reservoirs, where temperature and pressure are commonly very low. For instance, about 88% of heavy oil reserves in the northwest edge of Junggar basin is buried less than 600 m, where the formation pressure is commonly between 1.8 and 4.0 MPa and the formation temperature is 16° to 27°C.

1.1.2 Sedimentary Characteristics of Unconsolidated Sandstone Reservoirs

Sedimentary types of unconsolidated sandstone reservoir vary, including many facies of marine sediment and continental sediment, and most of them are continental facies, marine–continental transitional facies, delta facies, fluvial facies, and alluvial facies of fluviolacustrine transitional facies. There are also fan delta facies.

Lithology is mainly poor sorting and rounding fine, medium, and coarse sandstone. The components of sand particles are mainly composed of quartz, feldspar, and clastic debris with high clay content (commonly 6% to 9%).

In the Clear Water zone of Burnt Lake area of Cold Lake Oilfield, the thickness of the pay is 20 to 30 m and lithology is fine sandstone. Components of particles are 20% quartz, 20% feldspar, 60% clastic debris, and clay content of 10% to 20% of the total rock volume, mainly consisting illite, chlorite, smectite, and kaolinite.

The second zone of Dongying group of SZ 36-1 Oilfield in Bohai Bay is a set of delta front subfacies of near shore lacustrine facies, including underwater distributary channel, mouth bar, interdistributary area, distal bar, sheet sand, sedimentary microfacies, and meare subfacies. The properties of crude oil in the oilfield are high density, high viscosity, high content of colloid and asphaltene, low content of sulfur and wax, and low pouring point. The reservoir of SZ 36-1 is obviously a heavy oil reservoir.

1.1.3 Diagenetic Features of Unconsolidated Sandstone Reservoirs

Diagenesis refers to the physical and chemical functions when the deposits change to sedimentary rock and before the metamorphic process. Because the unconsolidated sandstone reservoir buries shallow, they generally experience compaction

Table 1-1 Sedimentary Facies and Tectonic Features of Typical Unconsolidated Sandstone Heavy Oil Reservoirs

Oilfield	*Formation*	*Sedimentary Facies*	*Structural Features*
Jinglou Oilfield	Sublayer III of Hetaoyuan Formation, Paleogene	Fan Delta	Anticline with complicated faults, asymmetrical wings
Gucheng Oilfield	Sublayer III of Hetaoyuan Formation, Paleogene	Fan Delta	Structural nose
Gudao Oilfield	Guantao Group, Neogene	River	With three main faults and several secondary faults
Qinhuangdao 32-6 Oilfield	Guantao Group, Minghuazhen Group, Neogene	River	Boundary controlled by two basement faults with five shallow secondary faults inside
Suizhong 36-1 Oilfield	Donging Group, Paleogene	Delta	Complex structure consisting of buried hill of pre-cenozoic, drape anticline of Paleogene, and fault nose
NB 35-2 Oilfield	Lower Minghuazhen, Guangtao, Neogene	Meandering river, braided river	Northeast strike complex nose structure consisting of traps of semi-anticline, fault blocks and north-south slope
Karamay Oilfield	Tugulu, Karamayi, Lower Wuhe, Fengcheng, etc.	Alluvial fan, fluvial	Located in northwest thrust belt of Junggar Basin, consisting of many fault blocks generated by a series of primary and secondary faults
Bell Creek Oilfield	Muddy sandstone	Mouth bar of fluvial facies	Anticline
Wilmington Oilfield	Tar, ranger, terminal etc.	—	Anticline with horsts and grabens destroyed by normal faults
Kern River Oilfield	Kern River and M grades	Continental alluvial fan	—
S.E. Pauls Valley Oilfield	Oil creek sand	—	Fault nose structure
Celtic Oilfield (Lloydminster)	Sparky, Upper Waseca of Mannville Group, carboniferous	Delta, fluvial, offshore sediment	Anticline
Cold Lake Oilfield	Cummings, Clear Water, and Sparky in Mannville Group, sarboniferous	Fluvial, dela, and prograding seashore	Anticline

Table 1-2 Relationship between Cement Content and Cementation Degree

Total Porous Space (%)	*Cement Content (%)*	*Porosity (%)*	*Cementation*	*Formation Features*
32	<5	27	Poor	Good
	5 to 14	18	Medium	Medium
	14 to 20	12	Fairly strong	Poor
	<24	8	Strong	Nonpay

and cementation at the early stage of diagenesis. To some extent, the rock strength depends on cementation of rock, the features of cements, the content of cements and cementation type. The poorer the cementation, the weaker is the rock. The cementing strength of rock depends on the components and contents of cements, the strength of rock cemented by clay is the poorest, that cemented by calcium is strong, and that cemented by silicon is the strongest. Table 1-2 shows the relationship between cement content and the level of cementation when the total interparticle void space is 32%. As seen from the data, the degree of cementation will increase with the incremental of cement content under the same conditions.

On the basis of the distribution of cements in the interparticle pores, the cementation types can be divided into basal cementation, porous cementation and contact cementation, etc., where basal cementation is the strongest, followed by porous cementation, and contact cementation is the poorest.

For unconsolidated sandstone reservoirs, they are commonly shallow and with weak cementation. Cements are mainly clay minerals, and calcium cements occurs occasionally. Major cementing types are contact cementation and pore cementation, whereas the rock with contact cementation is the weakest consolidation. There are some reservoirs without the function of cementation, where only sand particles can be cored by convention methods. For instance, the depth of the major pay of the lower zone of Minghuazhen group of NB 35-2 Oilfield is 1,100 m, and core samples are very loose. The samples will change to a pile of sand particles without the treatment of freezing. Table 1-3 shows the cementing features of some typical unconsolidated sandstone reservoirs.

1.1.4 Characteristics of Reservoir Space of Unconsolidated Sandstone Reservoirs

1.1.4.1 Type of pores

The types of pores formed in unconsolidated sandstone reservoirs are very rich through analysis of casting thin sections, images, and ambient scanning electromicroscopy (SEM), including mainly interparticle pores, microfracture,

Table 1-3 Cementing Situation of Some Unconsolidated Sandstone Heavy Oil Reservoirs

Oilfields	*Formation*	*Cementing Type*	*Cements*	*Cement Content (%)*
Gudao Oilfield	Guantao	Contact, pores Contact-pores	Mainly clay	9 to 12
QHD32-6 Oilfield	Minghuazhen Guantao	Contact	Mainly clay	<3
NB35-2 Oilfield	Lower Minghuazhen Guantao	Pores Contact-pores	Mainly clay	5 to 8
Tuyuke Oilfield	Karamay	Mainly contact	Mainly clay	6 to 7

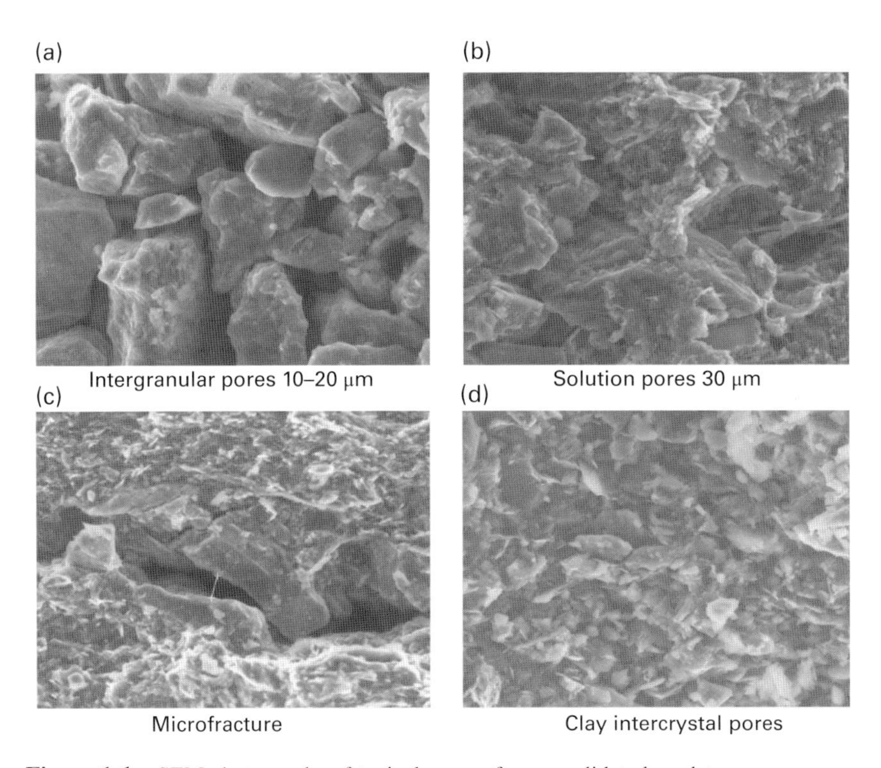

Figure 1-1 SEM photographs of typical pores of unconsolidated sandstone pay.

dissolution pores, dissolution fracture, and intercrystal pores (Figure 1-1). Generally, primary pores dominate the pore spaces and there are only a few secondary pores. The major pores are primary intercrystal pores, followed by the micropores of miscellaneous matrix, where siltstone also remains well-developed primary pores. For the secondary pores, most of them are the large dissolved pores, then intergranular solution pores and microfractures.

1. Primary intergranular pores
   Primary intergranular pores are developed fairly well in unconsolidated sandstone reservoirs, where the frequency of occurrence in the analysis samples of pore structure is close to 80%. Because the mechanical compaction of unconsolidated sandstone rock is weak, the primary intergranular pores are well preserved, mainly in siltstone, shaly siltstone, and sandy strips, where the media are mainly fluvial channels with good connectivity and the pores are open with the supporting of chemical sediments such as siderite. The diameters of pores are 3 to 60 μm, mainly in the range of 10 to 50 μm. The statistical areal porosity of samples is 0.5% to 22% and the average value is 4.55%. These pores are the main effective pores due to their large diameter, good connectivity, wide distribution, and so on.

2. Dissolved pores and fractures
   Dissolved pores and fractures are formed by the dissolution of unstable minerals during diagenesis, where the diameters of the pores are big, but the frequency of occurrence is less than that of primary pores. The maximum diameter of pores reaches 1 mm, mainly in the range of 50 to 500 μm, and the areal porosity based on the statistics of samples is 5%. Dissolved pores and fractures are mainly formed by epigenesis of dissolution. On the basis of SEM, these pores are mainly appearing in the buried depth of 900 m and less developed in the shallow formations. Though these pores are relatively rare, they play an important role in improving storage space due to their larger diameter.

3. Microfractures
   Microfractures exist in silty mudstone and shales with sandy strips. Their maximum width reaches 20 to 40 μm and extended length is up to 1 mm, providing good flowing channels by connecting with sandy strips. Most of them are diagenetic fractures, which are related with the sedimentary features of regions. The depositional speed is fast in unconsolidated sandstone formation, where the sedimentary particles are small and the drainage of water is not good during compaction. If the large segment shale occurs, the abnormal high pressure will form in these formations. The flow of water will form some sediments like carbonate and siderite to support micorfractures acting as drainage channels, which are mainly open and will be good permeable space provided that the further compaction is not occurring in formations.

4. Intercrystal pores
   Intercrystal pores mainly exist in shale of unconsolidated sandstone, and intercrystal pores of illite usually appear in the layer of illite/montmorillonite;

intercrystal pores are also observed in carbonate crystals. The diameter of intercrystal pores is generally small (in the range of 1 to 10 μm), and the frequency of appearance in formation is far less than that of intergranular pores.

1.1.4.2 Throat types

On the basis of observing SEM sections and casting slices, there are three types of throat in unconsolidated sandstone formations: pore narrowing throat, neck constriction throat, and tube bundle throat.

1. Pore narrowing throat
   These throats appear mainly in sandstone formations with intergranular pores; the throats are the narrowing parts and are difficult to identify them from pores. They are typical combination of large pores and big throats and are the main throats in good quality formations.

2. Neck constriction throat
   These throats are the constriction part of intergranular variable cross-section, mainly the residual constriction part of the rearrangement of rock particles under compaction, where most of them are the combination of large pores and thin throats. Because the mechanical compaction is not strong in unconsolidated sandstone formations, the size of throat is up to 5 μm and, therefore, the combination of large pores and medium-sized throats can be formed. These throats are more common in formations, sometimes coexisting with pore narrowing throats. With the increase of buried depth, the compaction will be strong and thus more neck constriction throats can be seen.

3. Tube bundle throat
   When the primary intergranular pores are clogged, most of the micropores within the miscellaneous matrix and cements are both pores and throats that are cross-distributed like lots of fine capillary tubes in miscellaneous matrix and cements, and therefore tube bundle throats are formed. These throats are usually seen in shaly siltstone and silty mudstone that have a high cement content or in a miscellaneous matrix, and they result from the combination of medium pores and fine throats.

1.1.4.3 Characteristic of pore structure

The parameters of pore structure can reflect comprehensively the size of pores and throats, distribution, and connectivity, so different parameters indicate varied features of pore structure. These characteristics influence each other, and there is a certain relationship among them. The research on the relationship between petrophysical features and parameters of pore structure in typical unconsolidated sandstone reservoir shows that there is no obvious relationship between porosity and permeability and other parameters of pore structure, while there is good correlation between permeability and parameters of pore structure. The correlation coefficient between permeability and parameters of pore structure like drainage pressure, maximum diameter of throats, median radius, median pressure, and so on is more than 0.8. Rocks with large median radius throat, large average diameter throat, and small drainage pressure will have higher permeability, and vice versa.

The correlation between permeability and skewness and kurtosis of the distribution of pores and throats is more obvious, and the correlation coefficient is as high as 0.9. The skewness and kurtosis are close to 0, and the permeability will be higher. The studies indicate that the distribution of throats is variable even when the porosity is close.

1.1.5 Fluid Properties of Unconsolidated Sandstone Reservoirs

1.1.5.1 High contents of colloid and asphaltene in heavy oil components

The main difference in components between heavy oil and conventional oil is the high contents of colloid and asphaltene in heavy oil and the low contents of oil. Generally, the contents of colloid and asphaltene in heavy oil are more than 30% to 50%, and the contents of alkanes and aromatics are less than 60% to 50% (Table 1-4).

Due to the fact that the components and composition of heavy oil and light oil are different, heavy oil has the features of higher viscosity and density. Based on the viscosity and density, heavy oil can be grouped into three types and further grouped into four types: conventional heavy oil (including I-1 and I-2), extra heavy oil, and ultra-heavy oil (or bitume) (Table 1-5).

Viscosity of heavy oil is extremely sensitive to temperature. Viscosity of heavy oil declines sharply with the increase of temperature, and the relationship between viscosity and temperature is linear in the ASTM coordination. When the temperature increases by 10°C, viscosity will usually be half of its original value. This is why we apply this thermal technology to produce heavy oil.

Table 1-4 Components of Heavy Oil

Country	*Oilfield*	*Viscosity (mPa·s)*	*Density (g/cm^3)*	*Component (%)*		
				Oil	*Colloid*	*Asphaltene*
China	SZ36-1	95.5 to176.3	0.968	—	11.16	9.83
	NB 35-2	201 to 741	0.968	—	21.85	6.33
Canada	Athabasca	—	1.015	43.49	23.39	18.0
	Cold Lake	—	0.994	53.57	28.32	15.0
	Peace River	—	1.026	50.00	30.50	19.5
Venezuela	Jobo	—	1.02	—	25.4	8.6

Table 1-5 Trial Standard of Heavy Oil Categories in China's Petroleum Industry

Categories of Heavy Oil			*Main Reference*	*Secondary Reference*
Name	*Category*		*Viscosity (mPa·s)*	*Specific Gravity (20°C, g/cm^3)*
Conventional Heavy Oil	I		50* (or 100) to 10,000	>0.9200
	Sub Category	I-1	50* to 150*	>0.9200
		I-2	150* to 10,000	>0.9200
Extra Heavy Oil	II		10,000 to 50,000	>0.9500
Extreme Heavy Oil	III		>50,000	>0.9800

* Viscosity of formation condition and other viscosity of dead oil under the condition of formation temperature.

1.1.5.2 Less paraffin and lower pour point with respect to heavy oil

The pour point of crude oil depends mainly on the content of paraffin, also on the content of heavy constituents in crude oil. The higher contents of paraffin, the lower is the pour point. Generally, the content of paraffin of heavy oil is less than 10% and its pour point is less than 20°C. Some fields in China have paraffin of less than 5%, and the pour point is lower than 0°C. For instance, paraffin content of heavy oil reservoirs in Karamay Oilfield is about 1.4% to 4.8%, and its pour point is –23 to –16°C. Paraffin content of heavy oil reservoir in Gudao Oilfield is 5% to 7%, and its pour point is –26 to –10°C.

1.1.5.3 Less gas and low saturation pressure with respect to crude oil

Heavy oil reservoirs contain less natural gas and light components as the biodegradation and damaging result in the loss of these components during the formation of these reservoirs. The light distillate oil content is low (low gas content) in these reservoirs. For example, light distillate oil content is generally lower than 10% at the temperature of 200°C, and original gas oil ration is generally less than 10 m^3/m^3 and sometimes less than 5 m^3/m^3. These reservoirs are with low saturation pressure and poor natural reservoir energy (Table 1-6).

1.1.6 Flowing Characteristics of Unconsolidated Sandstone Reservoirs

Mobility in the heavy oil reservoir decreases with the drop of reservoir pressure due to the especially high viscosity of heavy oil. The following flowing features of supervicious heavy oil are observed based on the laboratory core flowing curves (Figure 1-2).

1. The viscosity of heavy oil can be considered as constant and also the maximum value during the low range of pressure gradient. As a result, the flowing process of heavy oil in porous media is very slow and flowing velocity is very low.

Table 1-6 Original Reservoir Pressure and Solution Gas–Oil Ratio (GOR) of Some Heavy Oil Reservoirs

Oilfield	*Pressure (MPa)*			*Original GOR*
	Pr	*Pb*	*Pr-Pb*	*(m^3/m^3)*
Ninth Area, Karamay Oilfield	2.03	1.72	0.31	5.0
Jinglou Oilfield	3.09	1.42	1.67	5.1
Gucheng Oilfield	1.88 to 10.47	1.4 to 8.9	1.48 to 1.57	4.7 to 31.5
SZ 36-1 Oilfield	14.31	9.72 to 11.69	1.5	24 to 30
NB 35-2 Oilfield	10.5	4.5 to 10.48	2.19	10 to 24.3
Shanjiashi Oilfield	11.65	4.66	6.99	8.1
Gudao Oilfield	12.50	10.75	1.75	30.0
Gaosheng Oilfield	16.10	12.41	3.69	31.0

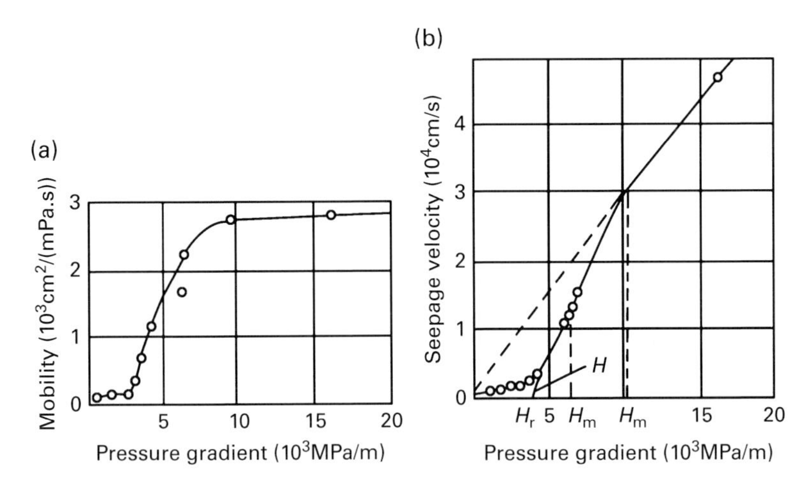

Figure 1-2 Curves of flowing rate of formation crude through sandstone samples with (a) mobility and (b) seepage velocity. (Cited from Sun Jianping, 2005.)

2. When the pressure gradient increases and reaches a critical value H_r, viscosity of heavy oil begins to decrease with the increase of pressure gradient and the mobility of oil increases accordingly. After the pressure gradient is over the critical pressure gradient value H_m, the flowing process will shift to the area of Darcy law, which means the flow will be constant with minimum viscosity of heavy oil.
3. An initial pressure gradient exists in the flowing of heavy oil in porous media, which is the pressure gradient when oil begins to flow in the bigger pores. With the increase of pressure gradient, medium and small pores are beginning to contribute to the flow.

Generally speaking, critical pressure gradient becomes lower for higher permeability zones. In high permeability zones with fast flowing velocity, pressure gradient of formation will be possibly greater than critical pressure gradient, even than the extreme pressure gradient. So the flow in high permeability zone will follow Darcy law. On the contrary, pressure gradient of low permeability formation coexisting with the high permeability zone will be lower than the critical pressure gradient, so part of the heavy oil will reside in low permeability pay. When producing in coexisting high permeability and low permeability pays, the effective viscosity (or mobility) difference between the pays will be dozens of, even hundreds of times and result in both imbalance of production process and uneven front conformance. In order to achieve better well performance in heavy oil reservoirs, it is necessary to increase pressure gradient to maximize large pores to contribute to oil production.

At the same time, measures should be taken to lower critical pressure gradient of heavy oil flowing and lessen the disadvantages of high viscosity of heavy oil. In practice, high production rate pumps are used in heavy oil reservoirs and generally high rate can be achieved in a short period. Actually, the formation pressure drops very fast by using this method and flowing velocity will decrease after pressure gradient of formation is lower than minimum pressure gradient H_m. Then, the production rate will decline slowly with very low production rate.

1.2 Development Technologies Management of Unconsolidated Sandstone Reservoirs

1.2.1 Development Characteristics of Unconsolidated Sandstone Reservoirs

Sand production is one of the major problems during production in unconsolidated sandstone reservoirs. Based on the sand production features observed during oil production, sand production is categorized into three types: unstable sand production, continuous sand production, and catastrophic high-rate sand production. The mechanism of sand production is very complicated, and sand production can occur during any processes, like drilling, production, or injection. Some factors resulting in sand production are due to engineering activities and thus can be avoided, while others cannot be controlled by operation. If the problem of sand production is not handled properly, sand production will be more and more serious and affect the development of oilfields and gasfields.

Sand production can cause a range of problems as follows:

1. Massive sand production will result in reservoir voidage, formation collapse, production decline, and so on. In the worst case, sand production can cause abandonment of wells.
2. The flowing of formation particles in unconsolidated sandstone due to flowing pressure, velocity and fluid viscosity can lead to disastrous consequences. For instance, sand production due to high flowing velocity or high sand concentration of fluid will easily cause abrasive corrosion of pipes, blocking pipes, and even production interval being buried by sand production.
3. The problem of handling massive amounts of produced sand in the wellsite on limited offshore platform and related environmental problems will occur.
4. The cost of artificial lift, gathering and transportation, supported facilities to dispose wastes, and so on will be increased.

The conventional heavy oil reservoirs in Bohai Bay, China have such features as relatively new formation, poorly cemented, unconsolidated, and low strength. Also, the viscosity of heavy oil is high, which will easily lead to sand production due to the strong drag force of fluid. In SZ 36-1 Oilfield, unconsolidated sand is observed from the cores. In this case, sand production is severe and causes the loss of production time after putting on production, such as sand burial of producing interval and sand blocking of pumps.

1.2.2 Technical Philosophy of Sand Production Management

Sand management is an integrated technology that enlarges the safely allowable sand production rate in practice, which aims at achieving the optimal production strategy between producing by excluding sand production and producing with sand based on optimizing production rate and productivity index of oil and gas wells. Sand management is an idea of design, decision making, and production management strategy based on comprehensively considering the factors such as well production rate, sand-carrying capability of wellbore, and surface-handling facilities and costs on the condition of producing by excluding sand production or producing with limited sand. This technology is different from the conventional technique of excluding sand definitely, and it is to achieve the optimal production management by controlling pressure drop of oil and gas wells, fluid-flowing velocity, and the inflow of formation sand. The idea of sand production management breaks through the traditional thought of excluding sand. Applications of controlled sand production in a few oilfields indicate that integrated sand production management can reduce cost of operation and increase well production rate. For unconsolidated sandstone reservoirs, sand production management is a new technology between excluding sand production and cold heavy oil production with sands (CHOPS). Well productivity index can be increased by improving porosity and permeability of pay on the basis of controlling sand production or producing with limited sand. With the development and improvement of geomechanics, mechanism and theory of sand production can be used to characterize sand production of pay properly, which will help to safely control the risk of sand production and maximize the overall benefits of operation. In addition, the research and development of new completion tools (e.g. oriented perforating tools) and improvement of well completion techniques are supporting the application of sand production management.

For potential sand production reservoirs, three techniques are applied: excluding sand, sand production management, and CHOPS. Table 1-7 compares the above-mentioned three development methods.

Table 1-7 Comparison of Excluding Sand, Sand Production Management, and CHOPS

Excluding Sand	*Sand Production Management*	*CHOPS*
Prevent sand production	Manage produced sand	Special requirements of formations
Lower production rate	Increase production rate	Produce massive amounts of sand
Reduced productivity index (PI)	PI increases with intermittent sand production	Difficult to control
Higher cost of sand control		
Less produced sand for handling	Need to handle produced sand	
Causing formation blocking	Formation blocking can be cured by itself	
Increase of skin factor	Decrease of skin factor	
Risk management not necessary	Risk management necessary	

1.2.3 Key Technologies

At present, general measures applied in unconsolidated sandstone reservoirs with sand production problems are to exclude sand production by using efficient well completion techniques. This approach will lead to the following consequences. First, production potential of wells cannot be fully fulfilled due to limited production drawdown. Second, measures to exclude sand production will increase skin factor and reduce well productivity index correspondingly. Third, if exclusion of sand production fails, oil and gas wells will be shut in, leading to the loss of production time and the increase of operating cost.

In recent years, with the development of the second phase of Suizhong 36-1 Oilfield and Qinhuangdao 32-6 Oilfield, gravel pack completion is widely used thanks to its better effect of controlling sand production and mature techniques. Because the mechanism of gravel pack is based on using packed gravel to prevent formation sand production and using screens to consolidate the packed gravels, a low permeability sand ring will be formed outside the screen inevitably and therefore production rate cannot be fulfilled due to increased skin factor. Especially for heavy oil reservoirs with low reservoir pressure, particle migration will be intensified with the development of oilfields and blocking will more likely be formed near to the wellbore. As such, flowing resistance of heavy oil from formation to wellbore will increase greatly, which results in significant drawback of oil production rate.

Aiming at improving the structure of porosity and permeability and keeping long-term oil production rate of wells stable as well as allowing fine sand particles to migrate freely and avoiding the destruction of skeleton sand, the development method of "sand production under control" draws increasing attention from the public. By referring to overseas oilfields producing with sand and laboratory and theory analyses, the conventional method of excluding sand production completely is updated and the idea of producing oil with limited sand production is presented, which will surely benefit the development of unconsolidated sandstone reservoirs in China and similar oilfields in other countries.

Sand production management is a complicated and systematic project, including different well completion types and parameters optimization of these completion techniques, prediction of sand production rate, monitoring technologies, artificial lifts that consider carrying produced sand, surface sand-handling techniques, and optimization of overall cost. Successful sand production management will cover the whole process from drilling to abandonment of wells.

Key technologies of sand production management are as follows:

1. What is the optimal sand production rate?
   - Formation skeleton damage, collapse of formation, and voidage will not occur as a result of sand production.
   - Sand production will form a disturbed zone with high porosity and further form a channel without definite geometry with increasing disturbed zone flowing with foam fluid. These channels are like fractures providing a drainage path with less resistance-wormhole network.
   - Handling capacity of surface facilities.
2. How is sand production rate controlled?
   - Select excluding sand production methods.
   - Optimize production parameters, control production rate, and drawdown on the basis of formation properties.
3. What technologies should be solved to achieve successful sand production management?
   - Prediction of sand production rate and monitoring techniques.
   - Relationship between oil production rate and sand production rate, and how to control sand production rate.
   - Analysis and calculation of sand-carrying capacity of well fluid.
   - Selection of well completion types and suitability study of artificial lift that considers limited sand production rate.
   - Additional sand-processing equipment and emissions without pollution.
   - Overall economic evaluation of producing with limited sand production rate.

1.2.4 Limited Sand Production Management

1.2.4.1 Proposal of limited sand production

With the increasing demand of petroleum worldwide, oil prices continue rocketing. Also, it is more difficult to find conventional light oil reservoirs. Under such situation, more and more countries and petroleum corporations are beginning to focus on heavy oil resources and trying to improve production rate of these resources. In order to enhance oil production rate, higher drawdown becomes necessary. In this case, migration of particles in the formation and sand production will occur in unconsolidated sandstone reservoirs. For a long time, the method of excluding sand production is adopted to produce oil. Excluding sand production is an important solution for the sand production problems; however, this technique will result in a high cost of well completion and low productivity of well, even the possibility of well shutdown. Although the techniques of excluding sand production are improved, actual well productivity is lowered to some degree.

If an appropriate technique that can both exclude a certain size of particles and allow limited sand production is adopted for unconsolidated sandstone heavy oil reservoirs, the cost of excluding sand production can be reduced to some extent, well productivity will be enhanced, and the development of this kind of reservoirs will be improved undoubtedly. For the above-mentioned purpose, it is proposed to produce heavy oil in unconsolidated sandstone reservoirs by allowing limited sand production. Pilot projects around the world indicate that to enlarge relationship between gravel size and formation particles and to control the size of producing sand will not block gravel packing interval and will instead improve well productivity.

To produce with limited sand production is to achieve optimal producing strategy between excluding sand production and allowing sand production on the basis of optimizing well production rate and well productivity. In fact, to produce with limited sand production is to exclude sand production selectively or limitedly. In other word, for sand production prone reservoirs, formation sand of different sizes will migrate with the flowing of heavy oil during production, so to purposely prevent the migration of certain and bigger size particles on the basis of analysis of movable sand sizes and distribution will build a sand production filter with the accumulation of migrated sand around the wellbore. Thereafter, smaller particles will be prevented from flowing with formation fluid. Finally, certain sizes of sand will be kept in the formation. Before the forming of sand production filter, flowing of smaller size of sand is allowed and formation properties near the wellbore can be improved, so formation potential can be fully used. Detailed techniques of this idea are stated next.

Based on the features of reservoirs, intermediate casing is installed and cemented above production interval, and premium screens (filtering size of premium screen is selected according to requirement of producing with limited sand production) are installed in open hole of the producing zone. After installing the excluding sand production equipment, electrical submersible pumps or electrical submersible progressive cavity pumps will be installed. To produce with limited sand production can be achieved by controlling drawdown after putting on production. Permeability at the wellbore can be enhanced a lot while producing with limited sand production, and well production rate can also be increased.

To produce with limited sand production is focusing on both excluding sand production and allowing producing with small amount of sand. Therefore, to handle well liquid and to clean up produced sand in time are very important in order to prevent sand from flowing into pipes and processors and affecting normal production when taking advantage of producing with limited sand production.

Producing with limited sand production is based on the detailed analysis of sand production mechanism, the quantitative relationship between sand production and well productivity, and optimization of excluding sand production. Therefore, the optimization of production of wells and oilfields can be achieved by using the best well completion and operating parameters. Some major features of producing with limited sand are summarized as follows:

1. Easy operation, less gravel pack tools and equipment, and reduced cost.
2. Less formation damage and better protection of producing zone, enhancement of oil production rate and recovery factor.
3. Preinstallation of surface sand separation equipment due to the volume of sand production at the beginning of putting on production and the time of water breakthrough to avoid the situation that produced sand cannot be treated in case of failure of gravel pack.

Model of producing with limited sand and processing diagram is shown in Figure 1-3 and Figure 1-4.

1.2.4.2 Reservoir conditions for the technique of producing with limited sands

Producing with limited sand in unconsolidated sandstone reservoirs on the basis of theoretical analysis and summarization of CHOPS is an extended development and application of CHOPS, which fully takes advantages of CHOPS, widens the application domain, and enables CHOPS to be used in

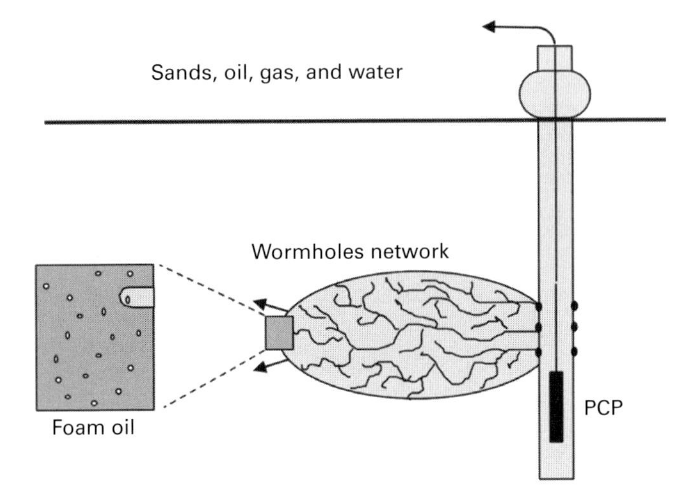

Figure 1-3 Diagram of producing process allowing limited sand production.

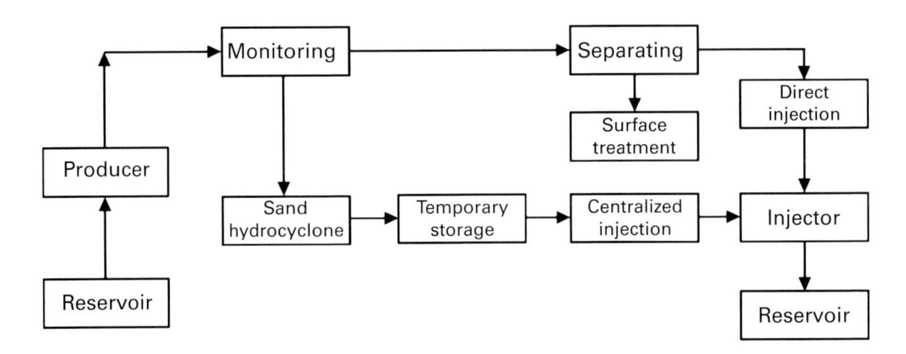

Figure 1-4 Treatment of produced sands.

more and more similar reservoirs. Based on prediction of conditions of sand production, sand production volume, and the quantitative relationship between sand production volume and well productivity, reasonable operating parameters, optimal strategy between excluding sand production and allowing limited sand production, and the selection of best well completion can be achieved. Generally, application of producing with limited sand is related closely with well completion, artificial lift, and supported equipment and facilities. For instance, if progressive cavity pumps or other artificial lift equipment cannot lift well liquid with sand from the bottom of wellbore to surface, the plan of producing with limited sand will not be feasible.

In order to apply successfully the technology of producing with sand in unconsolidated sandstone reservoirs, it is necessary to understand detailed characteristics of reservoirs and to screen the potential reservoirs. Because each reservoir is different from others, there will be no fixed rules to apply this technology for different reservoirs. When implementing the technology of producing with limited sand, the following aspects should be analyzed case by case, such as reservoir depth, formation thickness, petrophysics of pay, reservoir pressure, viscosity and density, original solution gas–oil ratio, and cementation of pay zone.

Generally, the conditions of application of producing with limited sand are similar with CHOPS. As a technology of controlling part of particles of pay zone, however, to produce with limited sand is somewhat different from CHOPS. Because research of producing with limited sand in unconsolidated sandstone reservoirs is a new domain and there is a lack of common results in many aspects, to screen potential reservoirs for the application of producing with limited still relies on past experience of CHOPS.

Although studies and experiments are in place in several research institutions and oil companies for many years, standards of screening potential heavy oilfields for CHOPS are not published officially. It is generally believed that the fields suitable for CHOPS should have the following features:

1. Unconsolidated sandstone reservoirs.
2. High porosity, permeability and oil saturation.
3. Reservoir depth is between 300 and 800 m.
4. Dead oil viscosity is between 2,000 and 40,000 mPa·s.
5. There is a certain amount of solution gas contained in oil (preferable value is about 10 m^3/m^3).
6. Low shale content in a pay.
7. Far enough from bottom water or edge water.

At present, well pattern of reservoirs producing with CHOPS is 160 to 400 m in general, while Professor Dusseault at University of Waterloo believes that reasonable well distance of CHOPS is about 170 to 400 m (i.e. 25 to 35 well counts/km^2).

To produce with limited sand aims at improving well productivity by allowing sand production, in other words, production rate of wells will be very low and reservoirs cannot be developed economically. So, in addition to considering the conditions of reservoir geology, pumps with high capacity of carrying sand, artificial lift technologies, gathering and treating equipment, etc. should also be prepared.

1.2.4.3 Applicable development phases of producing with limited sands

Theoretically, the technology of producing with limited sand in unconsolidated sandstone reservoirs can achieve the best result in the undeveloped reservoirs. Advantages of producing with limited sand can be fully achieved when integrated research, planning, designing, and execution are considered at the beginning. Integrated research on the basis of well type, reasonable well pattern, well completion, plan of matching facilities will bring the maximum potential of reservoir as well as save cost. Among several successful cases of applying technology of producing with limited sand, Husky Energy produced with sand by installing progressive cavity pump in Black Foot heavy oil reservoir (reservoir depth is 600 m and oil density is 0.97 to 0.98 g/cm^3) with low production rate and high water cut. As a result, in less than half a year, oil production rate was increased by 1 to 6 times and water cut reduced by 10% to 40% in this reservoir.

Field practices of producing with sand were implemented after steam huff and puff in heavy oil reservoir in Nanyang Oilfield, China. The field practice was not very successful because of the depletion of reservoir pressure and decrease of solution gas after several times of steam huff and puff. Therefore, the status of reservoir depletion should be considered when changing steam huff and puff to producing with sand.

1.2.4.4 Analysis of factors affecting development of producing with limited sand

1. Influence of sand production
   To produce with sand can greatly improve well productivity. Generally, sand concentration at the beginning of well effluent can be over 20%, which can result in the failure of normal production. But sand concentration will decrease to 0.5% to 3% after producing for 0.5 to 1 year with the production of oil and sand, and oil rate will increase continuously and tend to be stable.

2. Influence of solution gas
   At the beginning of putting on production, foam oil does not form because of high well bottom pressure. Foam oil begins to form with the drop of reservoir pressure after producing for some time. Oil film that covers gas bubble is strong due to high concentration of resin, asphaltene, so foam oil can keep stable for a long time. Actually, solution gas drive of foam oil is different from traditional solution gas drive. In general, recovery factor of reservoirs driven by foam oil and solution gas can reach 8% to 15% when producing with sand. Furthermore, foam oil

tends to more stable with faster pressure drop velocity, so to improve production rate a little bit will help foam oil to function better.

3. Influence of edge water, bottom water, gas cap, etc.
   As a reality, existence of bottom water and edge water will provide driving energy for production of reservoirs. In order to prevent formed wormholes from connecting with aquifer and avoid encroaching of water body, however, oil wells should be located in areas with huge barrier zone or far away from aquifer. Also, it is better to place oil wells in areas without gas cap. In addition, if there is water zone above the oil pay, perforation interval should be several meters far away from water zones to prevent wormholes from damaging cap rock and connecting water zone easily. If there is bottom water, perforation interval should be several meters above bottom water.

4. Influence of artificial lift pumps
   The production rate of wells producing with limited sand is generally high, so progressive cavity pump with high rate should be chosen and surface matching driving equipment should be adjusted to meet the increased torque. For wells with higher production rate, electrical submersible pumps should be used.

1.2.5 Measures of Controlling Sand Production

Unconsolidated sandstone reservoirs are widely spread in China and therefore sand production is a common problem during the process of oil production, which pushes techniques of controlling sand production forward. At present, there is a wide range of technologies to exclude sand production and new technologies in this domain are being adopted from time to time. Artificial sand production–controlling techniques can be categorized into the following types on the basis of features of excluding sand: artificial wellbore wall, screen and gravel pack, mechanical sand-filtering pipe, frac packing, and combination of mechanical sand-filtering pipe and chemical sand-controlling techniques. All these sand-controlling techniques are aiming at realizing high production rate, being effective in the long run, and developing oilfields in high level. Therefore, the following aspects should be satisfied when controlling sand production:

1. Can be realized in the pay zone safely.
2. Appropriate sizes of sand to be excluded should be determined with less impact on well productivity.

3. Screens should have strong resistance to pressure and deformation.
4. Resistance to corrosion should be reinforced if running for a long period of time.

1.2.5.1 Sand control by building artificial borehole wall

Applying an artificial borehole wall with precoated gravel is a technique that coats surface of quartz sand with a resin by physical and chemical method to form stable and cohesionless particles after drying in normal temperature. After bringing the precoated gravel to sand production pay by using sand-carrying liquid, resin on the surface of gravel will soften and consolidate in certain condition (it works by squeezing curing agent and temperature), artificial borehole wall with good permeability and strength will form, and the purpose of controlling sand production is finally realized. This technique is applicable in pays with good absorption capability and reservoir temperature higher than 60°C. This technique features easy installation and high ratio of success. The strength of resistance to pressure after consolidation of gravels can reach 5 MPa and the permeability of pays 90% of its original value.

Artificial borehole wall with cement sand mortar is a technique where cement (cohesion agent), quartz sand (proppant), and water are mixed at certain percentage, and carried by oil to downhole and outside of casing, accumulated in the sand production interval, forming artificial borehole wall with strength and permeability after consolidation to prevent sand production. The method is important in the late stage of sand control of oil wells, featuring high permeability, extensive sources of raw materials, and easy installation. The disadvantages are the need of high volume of oil, poor resistance strength of pressure (less than 1 MPa), and short-term effectiveness.

The major features of controlling sand by artificial borehole wall are as follows:

1. Applicable to formations with severe sand production problems and serious voidage.
2. Results of controlling sand production are decided by chemical that consolidates sand because artificial borehole wall is dependent on the chemical agent.
3. Chemical consolidating agent will damage the permeability of gravel zone, and the term of effective consolidation is difficult to identify.
4. Consolidation strength of chemical agent will be weakened with the development of oilfield, and sand production will occur. Especially for gas wells (with high flowing velocity) and oil wells with high production rate, sand production will be more severe.

1.2.5.2 Chemical sand-controlling techniques

Chemical sand control is to squeeze chemical agent solvent into sandstone formation by using diesel as pore-enlarging agent, then squeeze hydrochloric acid (consolidating agent) to consolidate unconsolidated sandstone under reservoir temperature to prevent sand production in wells. This technique is applicable to the early stage of sand controlling of oil and water wells. The strength of resistance to pressure after consolidation is about 0.8 MPa, permeability is retained by 50% of its original value, and resistance to temperature reaches 100°C. The other features include resistance to such medium as water, oil and HCl, intolerance to mud acid and easy installation. This technique, however, requires higher cost and longer execution.

Subsurface synthetic sand controlling with phenol and formaldehyde solvent is to squeeze phenol and formaldehyde and mix completely with catalyst and diesel (pore enlarge agent) by certain percentage to sand production pay. After squeezing into target formation, resin will be formed in the formation temperature and deposit on the surface of sand particles. Loose formation sand particles can be consolidated after taking effect of solvent. Diesel will not take action with other materials in this process as a continuous phase, which enables the consolidating formation to have good permeability, enhances consolidating strength of sandstone and prevents sand production.

The features of chemical sand-controlling method are as follows:

1. Applicable only to formations that do not produce sand severely.
2. Certain formation temperature is required.
3. Formation permeability in consolidating area will decrease by 50% and production rate will be reduced.
4. Only valid for a short term and greatly affected by high water cut.
5. Expensive chemical agents.

1.2.5.3 Sand exclusion by using mechanical filters

To exclude sand by using sand filters is a technique that suspends the sand-excluding filters in the sand production zone with the help of pipes and auxiliary tools, as shown in Figure 1-5. Sand filters are with high permeability, which allows formation fluid to pass through and prevents sand production. Formation fluid will flow through sand filter to wellbore and to surface finally. At present, commonly used sand filters include wire-wrapped screens, slotted liners, double layer prepaced gravel wire–wrapped screens, as well as a variety of sand filter and newly invented sand-filtering tools. These sand filters can prevent sand particles whose size is larger than slot width (or pores/mesh diameter) as well as the particles whose diameter is less than slot

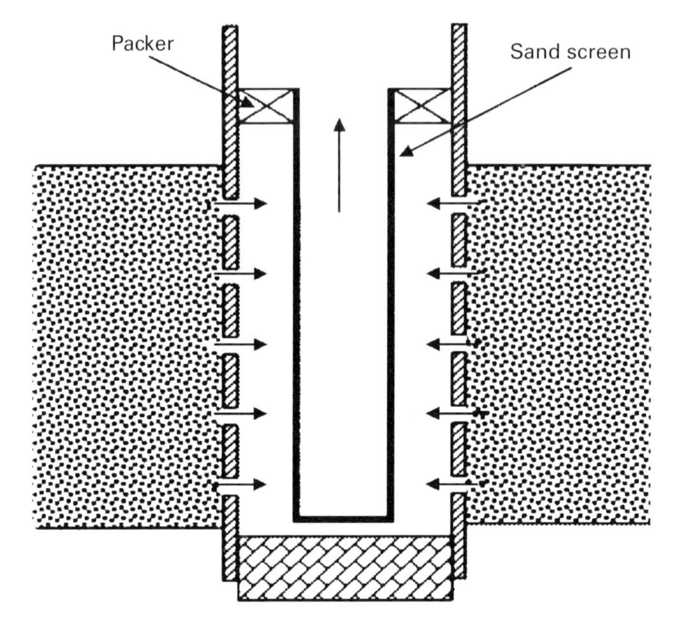

Figure 1-5 Diagram of sand exclusion by using sand filters.

diameter partly due to the "bridge" mechanism surrounding intervals from flowing into wellbore.

There are currently dozens of mechanical sand filters in use, including wire-wrapped screens, slotted liners, ceramic sand filters, metallurgy powder sand filters, resin quartz sand filters, composite microporous sand filters, metal (cotton/felt/mesh) sand filters, double-layer prepacked gravel, and wire-wrapped screens. Basically, the principle of sand filters is to utilize porous sand exclusion media formed by metal and gravel. Therefore, formation fluid can flow through sand filters freely, and sand particles from formation will be blocked selectively based on the designing accuracy of sand exclusion (Table 1-8).

The basic features of mechanical sand filters are as follows:

1. Cannot prevent further sand production of formation.
2. Well production rate declines after wellbore and annulus is filled with continuous sand production.
3. Sand filters can be blocked by clay and fine sands and therefore production will decline severely.
4. Easy installation, low cost, but only effective for a short term.
5. Not applicable for formations producing fine sand particles.
6. Not applicable for pays with severe problems of sand production.

Table 1-8 Structure and Function of Varied Sand Filters

No.	*Type of Sand Filter*	*Structure*	*Size (mm)*	*Functions*
1	Slotted liner	Slotted tubing or casing	0.3 to 0.6	Tailpipe for vertical well, slanted well, horizontal well, and multilateral well in formation composed of moderate to coarse sand particles.
2	Quartz sand filter	Base pipe plus outside pipe consolidated by quartz particles	0.06 to 0.15	Vertical wells. Easy to crush. Not applicable for offshore wells.
3	Single/ multiple layers wire-wrapped screen	Base pipe, longitudinal ribs, and wire wrapped outer layer	0.15 to 0.75	Tailpipe for vertical well, slanted well, horizontal well, and multilateral well, accompanying gravel packing sand control.
4	Double layers wire-wrapped and prepacked gravel screen	Base pipe, longitudinal ribs, wire wrapped layer, and wire wrapping outside resin consolidated gravel packing layer	Wire wrap seam 0.15 to 0.75; Gravel layer 0.06 to 0.15	Used independently or to support gravel packing sand control. Vertical well, slanted well, and multilateral well.
5	Double opening/ prepacked gravel screen	Base pipe, longitudinal ribs, wire wrapped layer, resin consolidated gravel packing layer, and outer protection steel pipe	0.06 to 0.15	Used independently in vertical well, slanted well, horizontal well, and multilateral well.
6	Wire braid sand screen	Base pipe, coarse metal wire net, fine metal wire net, and threaded pipe protective layer	0.01 to 0.1	Products of 1990s. Applicable for sand control in any wellbores of oil and gas wells.
7	Stainless particles sintered screen	Base pipe sintered with metal particles	0.01 to 0.08	Products of 1990s. Applicable for sand control in any wellbores of oil and gas wells.
8	Wire braided layer plus multiple layers sintered with metal particles screen	Base pipe, interbraided metal wires, sintered layer with multiple layers of fine net and metal particles, and outer threaded pipe protective layer	0.01 to 0.08	Products of 1990s. Applicable for sand control in any wellbores of oil and gas wells.

1.2.5.4 Gravel packing sand exclusion

Gravel packing sand exclusion is applied earlier with a longer history (Figure 1-6). With the improvement of theory, techniques, and equipment in recent years, gravel packing method is regarded as one of the most successful sand-controlling approaches in the meantime. Gravel packing sand-controlling technique accounts for nearly 90% sand-controlling jobs in overseas industry.

Screen and gravel packing sand controlling is to install wire-wrapped screen or slotted liner to sand production interval of formation, then build a layer of

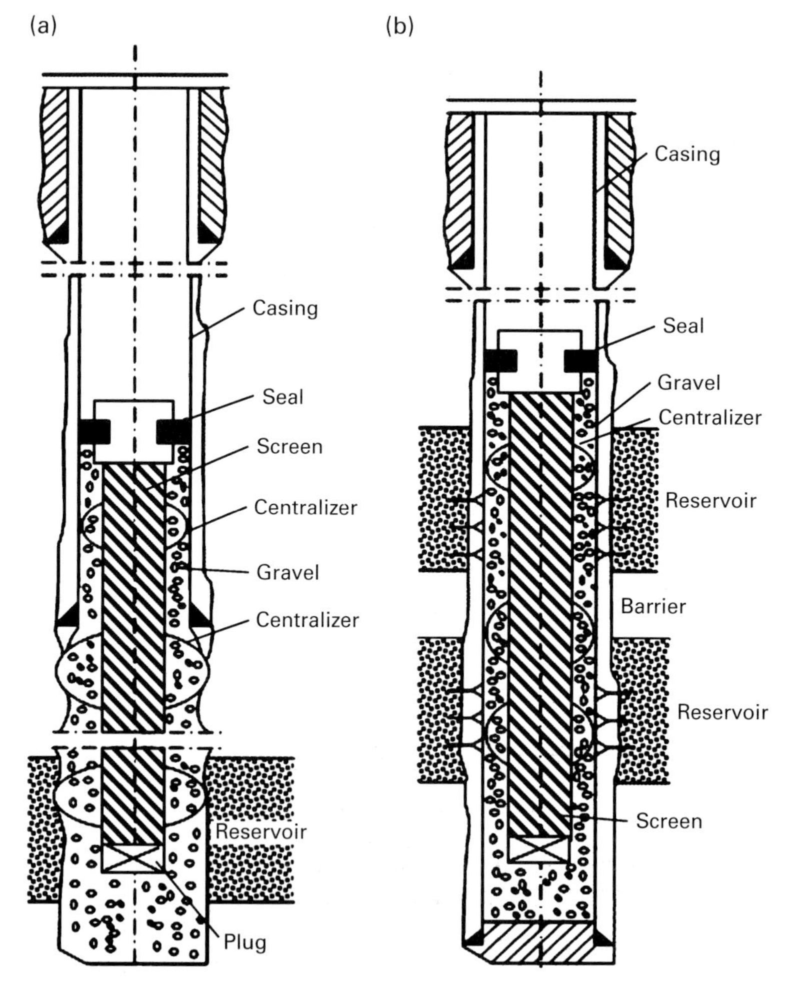

Figure 1-6 Diagram of screen and gravel packing sand controlling. (a) Gravel pack completion. (b) Gravel pack inside casing.

gravel between screen and tubing or casing by delivering selected gravels of certain sizes with carrying fluid to prevent formation sand from flowing into wellbore. Selection of gravel depends on the sizes of formation sand. The purpose is to keep formation sand flowing with oil outside the gravel barrier, and achieve both a good flowing ability and sand exclusion production by building naturally a sand arch from coarse to fine sizes outside the gravel barrier.

The major features of gravel packing sand control are as follows:

1. Screens are used to stabilize gravel layer, so gravel layer will be with good strength and will block sand production effectively.
2. Best method of controlling sand production with long-term effectiveness.
3. Applicable for formations producing fine size sand and those with massive amounts of sand.
4. High cost on both materials and operations.

1.2.6 Application of Sand Production Management

With theoretical development of geomechanics and technical improvement, more and more companies conducted studies and pilot testing on enhancing well productivity on the basis of sand production management of oilfields.

1.2.6.1 Laboratory experiments

In 1990s, corporations like Shell, Schlumberger, and TerraTek conducted simulation experiments of sand production and achieved lots of valuable results.

Schlumberger used a simple, easy to monitor experimental system to study sand production. In this study, an artificial core sample with perforation was selected, and factors like stress, flow volume, fluid viscosity, and rock strength were considered. These experiments showed that, for the same core samples, flowing velocity is the most important factor resulting in sand production. If the liquid is oil or water, there is a possibility that sand production will happen when the flowing velocity in perforated hole is over 0.2 m/s. Fluid flowing can be sorted into axial flowing and radial flowing, of which axial flowing (along axial of perforated hole) has a greater impact on sand production as it is easy to clean up broken rock particles on rock surface. So, it is suggested that the bigger diameter and high density perforation be used to improve critical pressure drawdown without sand production.

In 1992, Shell published a few results of large-scale laboratory experiments about sand production. The researchers used outcrop samples of a true formation in an oilfield as experimental core. The experiment was aiming at studying the influence of stress and flowing velocity on sand production. The size of experimental core used was 0.7 × 0.7 × 0.81 m. This experimental device can simulate three-dimensional stress, pore pressure, etc.

There are a few creative results in the experiments conducted by Shell, one of which shows that massive amounts of sand will be produced in the short term when drawdown increases, while sand production will keep at a low rate when drawdown is kept constant. It indicates that sand production is greatly included by variation of pressure. Furthermore, sand production rate increases when changing oil to salt water; then, sand production rate will reduce to the lowest level shortly.

Depending on various purposes, scales of simulation experiments of sand production are different. Research and development of some experimental equipment and experimental process are simple, while others are complex such as whole diameter big-scale sand production simulation experiments. However, the following aspects are all included in these experiments.

First, all experiments were aiming at reflecting the true situation of formation and building a set of complete, comparable technical parameters such as formation temperature, pressure, formation parameters, fluid medium, and formation strength. Second, the researchers all tried to improve accuracy of experimental devices like data collection, pressure monitoring, simulation of perforated hole, diameter and length of flooding cores, particle size analysis, computed tomography (CT) scan technology, and so on. For instance, to simulate dynamic changes of permeability, changes of sand production rate and fluid rate under actual condition of pressure drawdown and flowing rate, to obtain particle size distribution of artificial cores and its distribution after experiments by laser grain size analysis technology, and studies of the formation of wormhole network by CT sand technology, etc. were conducted. Third, theoretical models like liquid–solid coupling models were verified by using experimental results, providing theoretical foundation of oilfield dynamic analysis to achieve effective management and regulate the development of oilfield.

Based on previous sand production simulation experiments in this domain, the research center of China National Offshore Oil Corporation (CNOOC) researched and developed simulation experimental equipment to observe sand production characteristics on condition of simple sand control and conducted a lot of laboratory experiments. These works further discovered features of sand production and achieved a considerable amount of valuable results.

1.2.6.2 Wellsite pilot test

Oriented perforation technology was used in Varg Oilfield in the North Sea in order to control sand production rate within a tolerable range, and good results were achieved. Though pressure depletion happened after producing for

1.5 years, sand production rate was always within tolerable range. Compared with traditional well completion like gravel packing and sand-controlling screen, this technology saved costs and was successful in this case. Sand production volume was under control throughout life cycle of wells, so sand production problem was not severe in the late stage of this marginal field. After analyzing sensitivity of formation strength, strength of pays reduced significantly comparing with that of initial status, but sand production in great amount was not observed.

Statoil performed scientific management of formation sand production in Gullfaks offshore oilfield by controlling sand production rate within tolerable range by remote controlling technology, improving its well productivity by a large margin.

Integrated application of combining well completion and sand production management was conducted in Sawan gasfield, Pakistan, by increasing sand production rate to a tolerable range, where a high production rate of $2.83 \times 10^6 m^3/d$ was achieved. Compared with conventional cased-hole completion, this technology saved lots of cost and reduced risk of well completion failure.

Dulang Oilfield in Peninsular Malaysia is a field with complex structure. Because no sand-controlling measures were taken, a few problems occurred in this field such as sand production in big amounts, decrease of oil production rate, frequent shutting in wells and cleaning produced sand. High-volume sand production led to erosion of pipe walls, so production of wells had to be terminated or pressure drawdown had to be reduced, and platform was shut-down to perform well workover and pipeline sand cleanup. Eight years later (1991), research of well completion to improve well productivity was conducted, so sand production volume was controlled effectively, production rate was increased from 141 m^3/d to 1747 m^3/d and huge economic benefits were achieved based on reservoir simulation appraisal, residual oil distribution, and application of frac packing (Tip Screenout).

1.2.7 Development Prospects of Unconsolidated Sandstone Reservoirs

The heavy oil resources around the world are very rich, and reserves of heavy oil are far more than that of conventional light oil. In recent years, growth of reserves of light oil was slow and therefore production rate of light oil was limited. Under such situation, development of heavy oil effectively supplemented decreasing production rate of light oil. Heavy oil resource is also abundant in China; however, these resources were rarely used before 1982. With the development of science and technology in petroleum industry after 1982, development of heavy oil made new progress continuously. Newly found reserves were put into production, and the annual increasing ratio of heavy oil production has been kept above 30%. In 1995, production rate of heavy oil reached 1300×10^4 t and began to play an important role in the overall production rate in China.

2

Sand Production Mechanism and Changes of Rock Properties Affected by Sand Production

Research of sand production mechanism is the basis of sand production management. It is also a multidisciplinary and integrated work combing chemistry, geomechanics, and other disciplines. On the basis of previous achievements in sand production mechanism, this chapter analyzes mechanism and influencing factors of sand production, designs appraisal experiments of sand production on changing formation properties that combine the case of unconsolidated sandstone reservoirs in Bohai Bay, studies the mechanism of sand production from the standpoint of experiments, and explores the relationship between sand production and formation properties. The above-mentioned work builds a theoretical foundation for sand production management in unconsolidated sandstone reservoirs and provides a guideline for the selection of sand control techniques and its parameters.

2.1 Sand Production Mechanism and the Factors Affecting Sand Production

2.1.1 Mechanism of Sand Production

The reason for sand production by oil wells is that the original pressure balance around the wellbore is broken, resulting in yielding and original structure failure of formation rock. The mechanism of rock failure includes mechanical failure and chemical failure.

Sand Production Management for Unconsolidated Sandstone Reservoirs, First Edition.
Shouwei Zhou and Fujie Sun.

2.1.1.1 Mechanism of mechanical failure

There are three mechanical failure mechanism types of sand production for rocks.

1. Shear failure: The pore pressure will drop and effective stress will increase in the formation during production, so elastic deformation (in hard formation) and plastic deformation (in soft formation) will occur in rocks. The plastic zone will form in the disturbed formation and shear failure will occur when the plastic deformation reaches a certain level. When shear failure occurs, solid particles will be stripped from the rocks.
2. Tensile failure: When a sudden change in pressure exceeds the tensile strength of the formation, sand production will occur and the perforation tunnel will enlarge. Sand production will occur when effective stress around the wellbore is greater than the tensile strength of formation. In general, tensile failure occurs at the tip of the perforation tunnel and the wall of the perforation tunnel.
3. Cohesion failure: Cohesion failure mechanism is very important in weakly consolidated formations. Cohesion strength is a controlling factor that determines whether exposed surfaces will be eroded. These potential positions are perforation tunnels, wellbore surfaces of open holes, fracture surfaces of hydraulic fractures, shear planes, or other surfaces of boundaries. Cohesion force is related with cement and capillary forces. When drag force caused by flowing fluid is greater than cohesion strength of formation, sand production will occur in formations. Cohesion failure usually occurs in formations with low cohesion strength. In unconsolidated sandstone formations, cohesion strength is close to zero, and cohesion failure is the main reason of sand production.

2.1.1.2 Mechanism of chemical failure

Rock strength is controlled by the following factors: contact force and friction force between particles and cohesion force between particles and cement.

When a certain amount of movable water exists in formations and begins to flow, chemical reaction of water with other material will occur and dissolve parts of cement; therefore, rock strength will be reduced. Failure level of sandstone due to chemical reaction can only be estimated by detection of sandstone cement.

Illite swells and disperses easily after absorbing water, so it is sensitive to flowing speed and water. Illite/smectite mixed layer belongs to intermediates of smectite transformation to illite and is easy to disperse. Kaolinite lattice has weak cohesion force, so it is prone to particle migration and sensitivity of velocity.

Analysis of the content of clay minerals of unconsolidated sandstone samples from Bohai Bay shows that the major components of clay minerals are illite (46% in average), followed by illite/smectite mixed layer (17%) and chlorite (15%). Kaolilite is also a main component. This type of rock features weak consolidation and low critical flowing velocity of sand production. Further, water production will reduce rock strength and cause more serious sand production problems. Sensitivity of both velocity and water in unconsolidated sandstone reservoirs is severe, so the influence of sand production and velocity sensitivity on well productivity should be considered when predicting well performance and managing production of wells.

2.1.2 Factors Affecting Sand Production

Sand production is a very important issue when producing oil. In general, sand production will occur when the stress of particles is greater than the rock strength. Rock strength depends on cohesion force of cement, fluid adhesion, friction force between particles of formation, and gravity of the particle itself. Stress of particles includes tectonic stress, overburden pressure, drag force exerted on particles due to flowing of fluid, and pore pressure and force by production drawdown. As such, sand production is determined by various factors.

2.1.2.1 Tectonic stress

Tectonic stress near faults and the top of structures is greater and results in natural microcracks or joints due to the break of the internal skeleton of the original formation. Rock strength of these locations is weak, where sand production occurs easily and severely. For locations far from faults and at lower parts of structures, sand production is not serious. Therefore, more attention should be paid to locations with weak rock strength.

2.1.2.2 In situ stress

Pore fluid pressure reduces with continuous decline of pore pressure, leading to increase of effective stress of reservoirs and causing stress concentration at borehole wall, destruction of perforating tunnel, and movement of envelope. Pressure drop of formation can reduce the influence of tension on sand production, but shear failure in unconsolidated sandstone formations becomes more serious. Therefore, the original state of in situ stress and pore pressure in formations are the major factors affecting sand production.

Rock with the vertical and lateral in situ stress is in equilibrium before drilling. Vertical in situ stress depends on the depth of formation and average gravity of rocks, while lateral in situ stress is related to the depth of formations and characteristics of the rock (such as elastic, plastic, pore pressure, and so on). Original stress equilibrium near wellbore wall fails first when drilling, and rock will maintain maximum stress value during production process. Therefore, rocks around wellbore wall will fail first due to deformation.

Production in oilfields will definitely change the status of in situ stress, and formation fails when maximum stress of formation skeleton exceeds limits of strength. Once a small change of formation stress (such as gravitational stress, tectonic stress, fluid stress, thermal stress, and so on) occurs, strain of formation rock will follow Hooke's law. In this case, changes in permeability and porosity result from applied load and are reversible. When formation pressure declines continuously and reaches critical value such that the load acting on rock is equal to the rock elastic limit, sand production occurs (although it is not severe). With the development of oilfields and gas fields, formation stress will be less than the elastic limit of rock, and sand production in great amounts will occur due to failure of the sandstone.

If the water injection rate cannot meet the requirements of the water injection plan during the development of oilfields, formation pressure will drop continuously and rock skeleton of pay will endure an increasing load. Formation skeleton will fail and expand outside on the condition of the loss of in situ stress balance. Sand production will worsen with expansion of the sand-producing zone.

2.1.2.3 Rock strength

Rock strength of formation indicates degree of cementation of rock particles of pay and is a major factor affecting sand production. In general, rocks with uniaxial compressive strength of less than 7.0 MPa are regarded as weakly consolidated rocks as well as a potential sand-producing zone. Cementation of formation is related to formation depth, type and quantity of cements, cementing mode, particle size, and so on. The features and quantity of cements play an important role in rock strength. Sand production due to failure of rock in nature means that cementation is broken, forming discrete sand particles. Failure of cement is affected by shear failure and extension failure, as well as dissolution influence of liquid. This is why some wells do not produce sand before acidizing, water breakthrough, or water injection but produce sand thereafter.

In general, the shallower the formation, the weaker is the compaction and the lower is rock strength. Sandstone cemented by clay will be very loose and have less strength. In addition, the performance of clay cements is unstable and easily affected by external interference, so cementing quality is affected.

2.1.2.4 Heterogeneity of formation

The influence of heterogeneity includes two major aspects: in situ stress and permeability. In site stress covers structural stress, overburden pressure, drag force on rock due to fluid flowing, pore pressure of formation, and production drawdown. Different rocks have various stress effects and different failure modes under the same load due to factors like diagenetic environments and developmental conditions. For weakly consolidated sandstone reservoirs, stress heterogeneity results in shear yield of formation in some directions easier than other directions, where the original structure of pay will fail and sand production will occur.

2.1.2.5 Production drawdown

Pressure of overburdened rock is balanced by fluid pressure of pores and the inherent strength of rock itself. The higher the production drawdown, the lower is the fluid pressure of pores. Therefore, effective stress on the rock particles will be greater and support of sand arching near wellbore will be weaker. Rock skeleton will fail when the rock strength limit is exceeded. The erosion force and drag force on rock due to fluid flowing in the formation are major causes of sand production, which means that higher production drawdown will increase the risk of sand production.

In addition, with some sudden changes like production drawdown, operating parameters of gas wells will cause or worsen sand production because of the changes in stress conditions of formation rock.

2.1.2.6 Water production

For unconsolidated sandstones that are weakly cemented, poorly compacted, and have a high content of clay, when water breaks through in the formation and makes the clay wet, the cohesion force between particles will be reduced, weakening rock strength and changing consolidated sands into loose sands. Therefore, after water breaks through, critical flowing velocity, which makes sand production occur, will be lower and sand production becomes easier.

To maintain formation energy and prevent pressure from dropping during the development of oilfields, water injection is commonly applied. Water injection will make clay in formation expand and migrate, so cohesion of clay is influenced severely. This theory explains why sand production will not occur when water cut of wells is zero and why massive sand production occurs after water breaks through.

2.1.2.7 Fluid viscosity

Drag force of flowing fluid acting on rock is greater when viscosity is higher, so rocks around wellbore produce sand easily due to failure of extension. Also, this is the reason sand production occurs easily in heavy oil reservoirs.

2.2 Changes of Formation Properties due to Sand Production

2.2.1 Experiments Observing Influence of Sand Production on Permeability

The purpose of this experiment is to evaluate permeability changes when different sizes of particles are drained quantitatively and in a step-by-step manner. It mainly focuses on the influence of sand size, the sand production rate, and initial permeability on formation permeability when allowing sand production, and it provides controlled sand size for sand production management.

Laboratory experiments can apply either natural cores or artificial cores, as sampling and processing unconsolidated sandstone are very difficult and thus there are almost no complete natural cores in these reservoirs. Cores used in the experiments are prepared by formation sand, and the sand tube used is the cylindrical stainless steel pipe (Figure 2-1).

2.2.1.1 Influence of sand production on permeability

Experimental methods and processes are as follows:

1. Fill sand tubes, and weigh and test initial permeability.
2. Reduce weight of formation sand in step 1 to simulate sand production, and keep mass fraction of sand production at 0.5%, 1%, and 2%, respectively.
3. Retest permeability.
4. Calculate results of permeability change.

Experimental results are shown in Figure 2-2.

Figure 2-2 shows that (1) With the increase in sand production, permeability increases nonlinearly. When mass fraction of producing sand is 0.5%, permeability is increased by 0.5% to 6.9% (3.7% average). When mass fraction of producing sand is 1.0%, permeability is increased by 2.2% to 21.4% (9.3% average). When mass fraction of producing sand is 2.0%, permeability is increased by 10.8% to 53.1% (25.2% average). (2) The percentage of permeability increase in different core samples varies greatly, possibly related to the initial particle size distribution of rocks.

2.2.1.2 Influence of particle size on formation permeability

The purpose of this experiment is to appraise permeability change after producing different ranges of particle size quantitatively.

Figure 2-1 Experiment equipment to observe the influence of sand production on permeability.

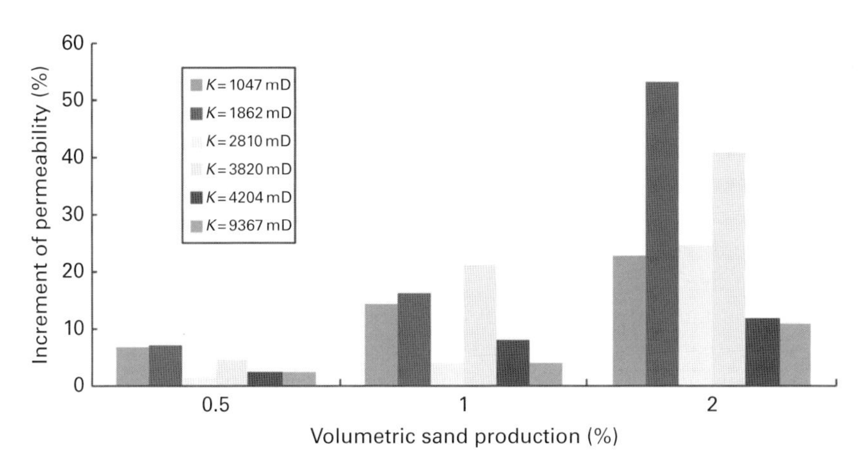

Figure 2-2 Influence of different sand production rates on permeability.

Experimental methods and processes are as follows:

1. Wash oil from the cores and dry.
2. Fill sand tubes, weigh, and test permeability using nitrogen.
3. Use standard mesh sieve to prepare particles with different size ranges, and conduct experiments based on producing 1% and 2% of mass fraction of sand, respectively.

4. Retest permeability.
5. Calculate results of permeability change.

Experimental data indicate that permeability of sand tube increases after producing sand (Table 2-1, Figures 2-3 and 2-4).

The following results are achieved on the basis of experiments:

1. Permeability of cores increases by about 20% when producing 1% sand of different sizes. Actually, sand production particle sizes ranging from 26 μm to 39 μm influence permeability greatly, and the average increase of permeability is 27.97%. Relatively large skeleton particles have less increase in permeability; for instance, sand production particle sizes ranging from 154 to 200 μm result in a permeability increase of 14.99% (Figure 2-3). The reason for these results is that formation particles less than 39 μm distribute around the pore-throat of rock dispersedly, which is the particle source of bridge blocking pore-throat and easy to migrate. Therefore, permeability changes greatly after producing this part of particles.
2. For cores with lower initial permeability, permeability increases greatly after producing sand. For instance, average permeability increases by 29.17% after producing 1% sand for samples with permeability of 431 mD. When permeability is more than 1500 mD, initial permeability will not obviously influence the relationship between permeability increase and the sand production rate.
3. The more sand production, the more increase in permeability occurs. Figure 2-5 compares permeability changes of sand samples from two different wells after producing different amounts of sand. For particles whose size is less than 50 μm, permeability increases more than one time and the amount of produced sand doubles. For particles whose size is bigger than 50 μm, increased rate of permeability is less than that of increased produced sand. In general, the bigger the particle size, the less permeability changes.

2.2.1.3 Evaluation of influence of producing sand gradually on formation permeability

The purpose of this experiment is to appraise permeability change after producing formation particles with different size ranges. We produce sand of different ranges of particle sizes gradually by using the sieve analysis method and then measure permeability after producing sand particles of 22, 25.8, 38.5, 42.0, 49, 61, and 74 μm, respectively. The following results were achieved (Table 2-2).

Table 2-1 Permeability Change after Producing 1% Sand of Different Size Ranges

		Permeability Change (%)							
Well Number	*Depth (m)*	*22 to 26 μm*	*26 to 39 μm*	*39 to 50 μm*	*50 to 74 μm*	*74 to 100 μm*	*100 to 154 μm*	*154 to 200 μm*	*Initial Perm. (mD)*
NB35-2-2	991.60	23.14	25.19	15.57	11.24	9.62	17.88	12.09	2102
NB35-2-2	991.90	22.99	31.72	25.20	18.42	17.90	22.92	16.68	1423
NB35-2-2	1000.90	35.81	37.60	29.53	24.10	23.49	24.47	—	431
NB35-2-4	1114.11	28.85	31.45	19.17	15.64	14.88	20.58	17.79	2093
NB35-2-4	1116.11	23.25	28.90	16.02	14.66	13.37	15.21	12.80	4147
NB35-2-5	1001.80	28.10	32.76	23.15	19.06	15.45	17.58	16.43	6285
NB35-2-5	1006.80	20.81	26.20	19.07	16.43	10.16	11.97	—	5562
NB35-2-5	1016.30	25.86	28.79	21.99	18.29	15.23	20.39	19.05	5445
NB35-2-5	1043.90	17.90	23.52	17.92	16.87	15.06	17.71	13.10	5439
NB35-2-6	1076.90	16.17	23.05	19.15	14.57	10.27	16.87	13.97	8441
NB35-2-6	1080.00	20.22	25.39	16.92	15.70	13.31	17.14	14.08	1815
NB35-2-6	1102.70	17.55	21.07	17.30	15.43	12.65	15.96	13.89	2982
Average		23.39	27.97	20.08	16.70	14.28	18.22	14.99	—

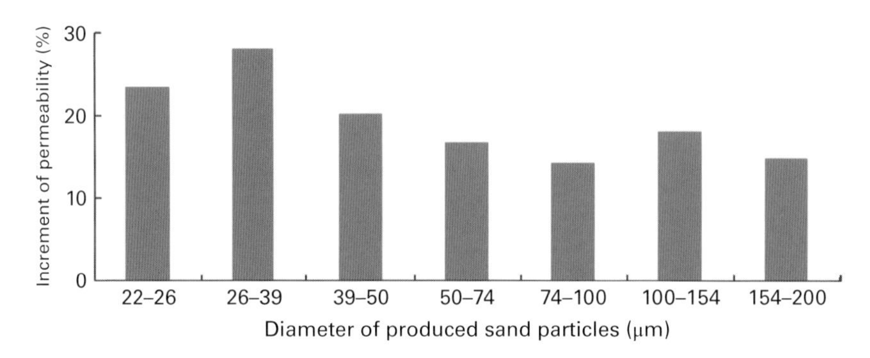

Figure 2-3 Average value of permeability change after producing sands of 1% w/w of different sizes.

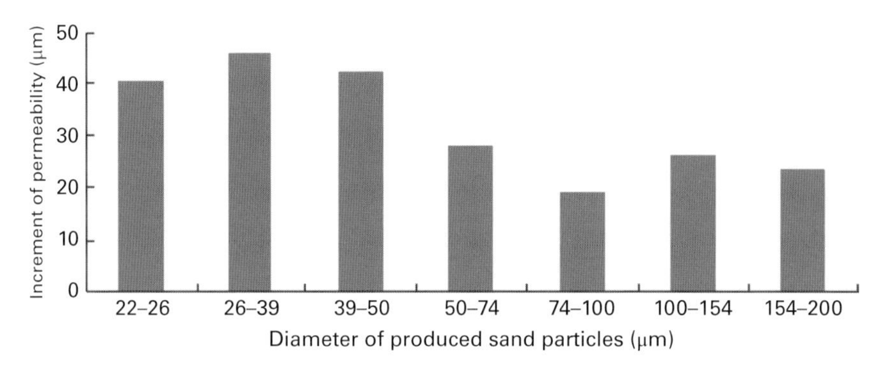

Figure 2-4 Average value of permeability change after producing sands of 2% w/w of different sized particles.

1. Permeability increases gradually with the increase in amount of sand production.
2. For particle sizes less than 22 μm, the sand production rate is 0.4% to 0.7% (0.14% average), while permeability increases by 5% to 22% (12% average).
3. For particle sizes between 22 and 25.8 μm, the sand production rate is 1.0% to 4.73% (2.17% average), while permeability increases by 20% to 125% (55% average).
4. For particle sizes between 25.8 and 35.8 μm, the sand production rate is 0.72% to 9.96% (5.37% average), while permeability increases by 14% to 780% (276% average).
5. For particle sizes between 35.8 and 42 μm, the sand production rate is 0.07% to 0.35% (0.19% average), while permeability increases by 1% to 16% (7% average).

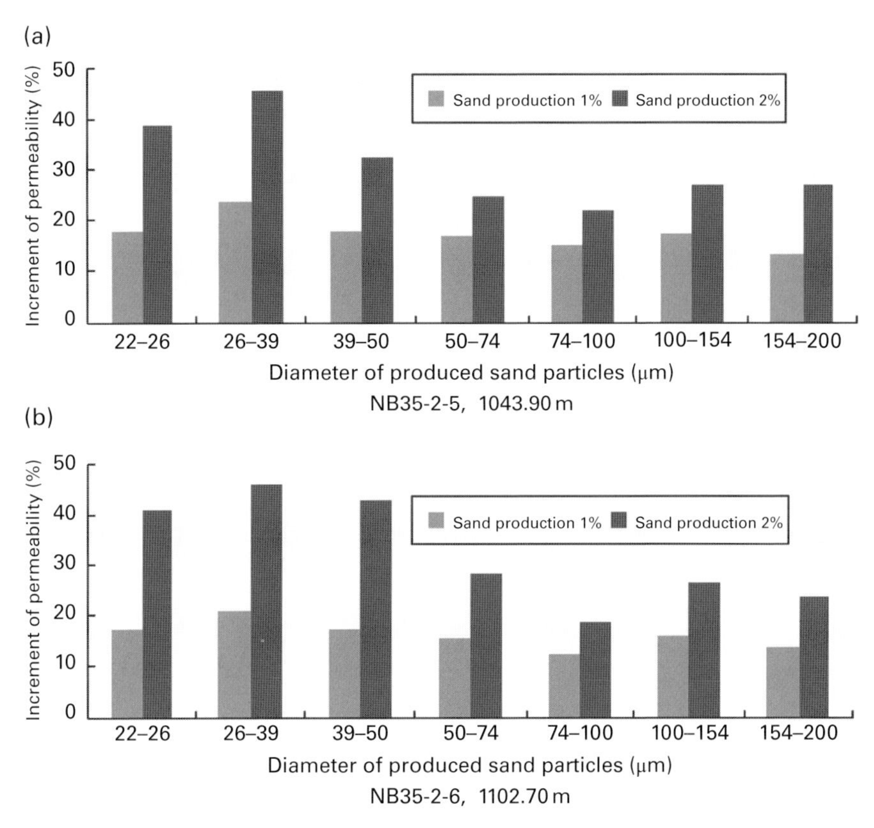

Figure 2-5 Comparison of permeability change after producing different sizes of sands.

6. For particle sizes between 42 and 49 μm, the sand production rate is 0.12% to 0.89% (0.49% average), while permeability increases by 3% to 23% (10% average).
7. For particle sizes between 49 and 61 μm, the sand production rate is 2.57% to 7.20% (4.36% average), while permeability increases by 20% to 57% (38% average).
8. For particle sizes between 61 and 74 μm, the sand production rate is 1.32% to 5.03% (2.23% average), while permeability increases by 5% to 25% (15% average).

In general, permeability will increase greatly after producing sand particles whose size is less than 38.5 μm. In other words, if the particles less than 38.5 μm are all produced, the sand production rate is 8%, and porosity will increase by 5.5% and permeability will increase more than 3 times (Figure 2-6).

Table 2-2 Statistics of Influence of Cumulative Sand Production of Different Particle Sizes on Permeability

Well	*Depth (m)*	*Parameter*	*Different Size of Produced Sand Particles*							
			Initial	*<22 μm*	*<25.8 μm*	*<38.5 μm*	*<42 μm*	*<49 μm*	*<61 μm*	*<74 μm*
NB35-2-5	1016.3	Cumulative produced sands (%)	0.00	0.44	5.17	6.68	6.79	7.07	10.24	11.66
		Perm. (mD)	7517	8524	19,146	23,743	25,024	26,711	33,758	38,149
		Increment of perm. (%)	0	13.39	154.71	215.86	232.90	255.34	349.09	407.50
NB35-2-2	1000.9	Cumulative produced sands (%)	0.00	0.64	2.33	12.29	12.53	13.42	19.30	23.29
		Perm. (mD)	1377	1473	2326	20,472	21,771	26,728	40,536	50,767
		Increment of perm. (%)	0	7.00	68.94	1386.69	1481. 07	1841.06	2843.79	3586.75
NB35-2-6	1080	Cumulative produced sands (%)	0.00	0.32	1.51	8.17	8.38	8.92	13.43	15.65
		Perm. (mD)	3868	4207	5317	22,190	23,776	27,039	35,424	40,295
		Increment of perm. (%)	0	8.76	37.46	473.67	514.69	599.05	815.81	941.76
NB35-2-2	991.9	Cumulative produced sands (%)	0.00	0.38	1.43	9.84	10.06	10.68	15.65	17.93
		Perm. (mD)	2686	3022	4310	21,287	24,707	27,278	38,083	45,020
		Increment of perm. (%)	0	12.52	60.46	692.50	819.83	915.57	1317.85	1576.09

NB35-2-4	1114.1	Cumulative produced sands (%)	0.00	0.14	1.84	9.34	9.65	10.40	17.60	22.63
		Perm. (mD)	3889	4155	4993	17,798	19,649	22,103	34,799	41,212
		Increment of perm. (%)	0	6.83	28.38	357.65	405.24	468.34	794.81	959.70
NB35-2-5	1006.8	Cumulative produced sands (%)	0.00	0.61	4.35	5.29	5.42	5.60	8.61	10.33
		Perm. (mD)	9210	10,592	21,581	24,976	26,429	27,382	35,898	40,770
		Increment of perm. (%)	0	15.00	134.32	171.18	186.96	197.30	289.78	342.67
NB35-2-6	1076.9	Cumulative produced sands (%)	0.00	0.72	4.78	5.50	5.57	5.69	8.26	9.97
		Perm. (mD)	9710	11,770	24,469	27,908	28,098	28,863	34,690	37,273
		Increment of perm. (%)	0	21.21	152.00	187.41	189.37	197.25	257.26	283.91
NB35-2-2	991.6	Cumulative produced sands (%)	0.00	0.72	1.72	10.22	10.40	11.04	14.58	16.11
		Perm. (mD)	2405	2933	4021	26,025	28,503	30,125	42,313	45,341
		Increment of perm. (%)	0	21.95	67.17	982.12	1085.16	1152.60	1659.38	1785.26

(continued)

Table 2-2 *(Continued)*

Well	*Depth (m)*	*Parameter*	*Different Size of Produced Sand Particles*							
			Initial	*<22 μm*	*<25.8 μm*	*<38.5 μm*	*<42 μm*	*<49 μm*	*<61 μm*	*<74 μm*
NB35-2-5	1001.8	Cumulative produced sands (%)	0.00	0.48	2.52	5.61	5.72	6.07	10.38	11.84
		Perm. (mD)	9108	9994	14,584	26,860	28,033	29,921	41,875	46,591
		Increment of perm. (%)	0	9.73	60.13	194.92	207.80	228.53	359.79	411.57
NB35-2-4	1116.1	Cumulative produced sands (%)	0.00	0.53	2.26	9.28	9.55	10.17	16.59	20.33
		Perm. (mD)	4154	4443	5661	19,780	21,165	23,424	33,049	39,480
		Increment of perm. (%)	0	6.95	36.27	376.17	409.51	463.88	695.58	850.40
NB35-2-5	1042.9	Cumulative produced sands (%)	0.00	0.56	2.05	5.73	5.88	6.17	9.84	11.16
		Perm. (mD)	8600	9922	12,664	26,240	26,611	28,275	38,498	43,203
		Increment of perm. (%)	0	15.37	47.26	205.12	209.43	228.78	347.66	402.36
NB35-2-6	1102.7	Cumulative produced sands (%)	0.00	0.47	1.65	9.69	10.04	10.71	15.84	17.73
		Perm. (mD)	2600	2739	3644	22,783	24,111	27,294	38,294	44,443
		Increment of perm. (%)	0	5.33	40.15	776.28	827.33	949.77	1372.83	1609.35

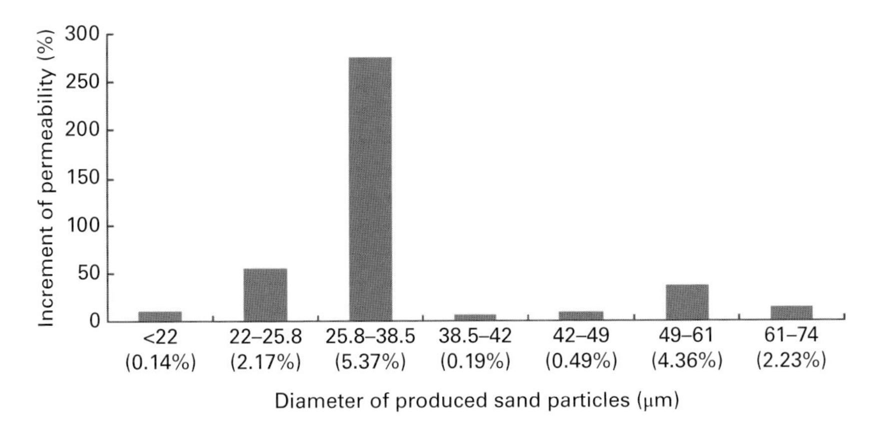

Figure 2-6 Influence of sand production of different particle sizes on permeability changes.

The formula for calculating increase of porosity resulting from massive sand production after a series of derivation is as follows:

$$\Delta\phi = \eta \times (1 - \phi_i) \times 100\%$$

where $\Delta\phi$ refers to porosity increment (%), η refers to sand production weight ratio (%), and ϕ_i refers to initial porosity of samples (%).

Laboratory experiments revealed that when the sand production rate reaches 5.61% to 12.29%, porosity will increase by 4% to 8% and rock samples will be weakly consolidated. The maximum sand production rate occurs in the rock at this time, and channel of sand production begins to form accordingly.

The original porosity of experimental rock samples of the NB35-2 oilfield is 0.33. After producing particles whose size is less than 38.5 μm and whose mass content is 8.14%, porosity of samples increases by 5.45% (its total porosity is 38.45% and permeability increases three-fold). On the basis of the relationship between porosity and permeability of natural cores, increase of porosity will lead to an exponential increase in permeability when porosity is greater than 35%. Our results of experimental evaluation coincide with this conclusion.

2.2.2 *Influence of Sand Production on Porosity Structure*

The purpose of this experimental study is to further appraise distribution features of a pore structure before and after the dynamic experiments by changing injection velocity, oil viscosity, production drawdown, and so on when conducting experiments that allow limited sand production or exclude sand production. In other words, by observing the characteristics of distribution changes of throats connecting pores of the same sample before and after migration of particles, especially the distribution characteristics of major throats, the size of the throat that is blocked during sand production can be identified. Therefore, reasonable parameters of controlling sand production can be achieved by combining the distribution features of sand production particles and the particle size that easily blocks throats on the basis of the corresponding initial drawdown, flowing velocity, oil viscosity, and so on.

Pore structure parameters of rock samples of sand tube in the experiment are measured by capillary flowing pore structure analyzer. The overall experimental equipment is shown in Figure 2-7.

One of the advantages of a capillary flowing pore structure analyzer in analyzing pore size of rocks is that comparison of the pore structure parameters of the same sample before and after conducting experiments is effective, accurate, and valuable. Moreover, the experimental medium is nonwetting nitrogen, and the wetting medium is the formation water or KCl solution, which will not change interface properties inside pores and will not affect future appraisal experiments. In comparison, the conventional mercury testing is destructive to core samples with poor reference values due to the strong microheterogeneity of rocks.

Figure 2-7 Experimental equipment of analyzing influence of sand production on pore structure.

The capillary flowing pore structure analyzer is used in the following aspects of appraising microaspects of formation:

1. To determine starting pressure and corresponding maximum effective throat after working fluid flows into rock.
2. To conduct testing for comparison analysis of pore size distribution of the same rock samples before and after acidizing, conformance modifying, compatibility experiment, various sensitivity experiments, dynamic damage experiment of drilling fluid, chemical flooding, and so on.
3. To identify effective pore size distribution before and after experiments such as allowing limited sand production or excluding sand production.

2.2.2.1 Analyzing the influence of pore structure parameters on permeability

Statistics on pore structure parameters of more than 150 samples were collected, and the relationship between these parameters and permeability was analyzed. Statistical results are shown in Figures 2-8 to 2-13.

Figure 2-8 is the relationship between porosity and permeability, and shows that the influence of porosity on permeability is not obvious when porosity is less than 30%. When porosity is between 30% and 35%, permeability will increase a little bit, but the increasing ratio is not big. When porosity is larger than 35%, permeability will increase obviously with the increase of porosity. Especially, permeability will increase exponentially with the increase of porosity when porosity is larger than 40%.

Figures 2-9 to 2-11 show the relationship between permeability and maximum pore-throat radius, median radius, and average pore-throat radius,

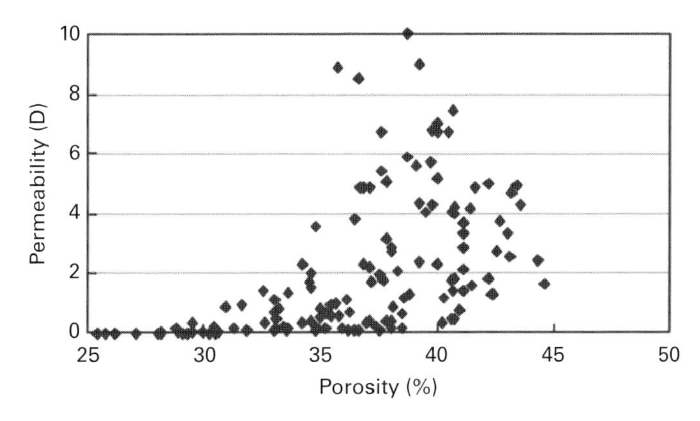

Figure 2-8 Relationship between permeability and porosity.

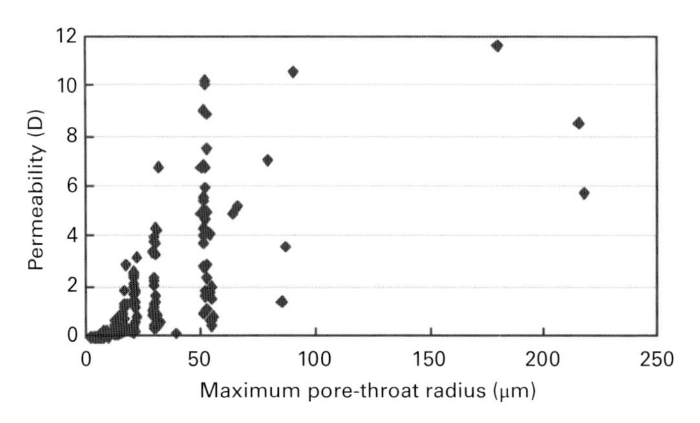

Figure 2-9 Relationship between permeability and maximum pore-throat radius.

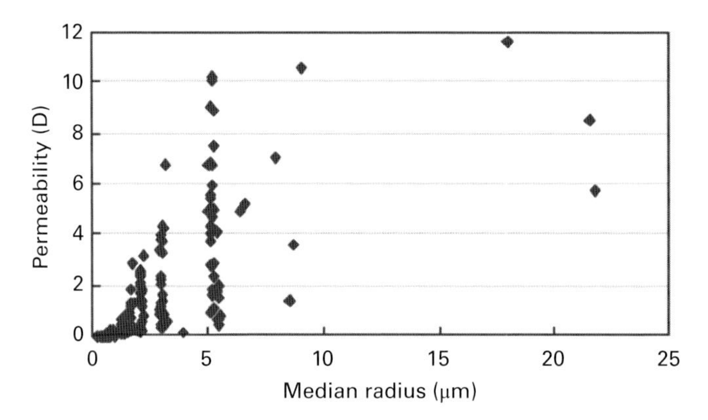

Figure 2-10 Relationship between permeability and median radius.

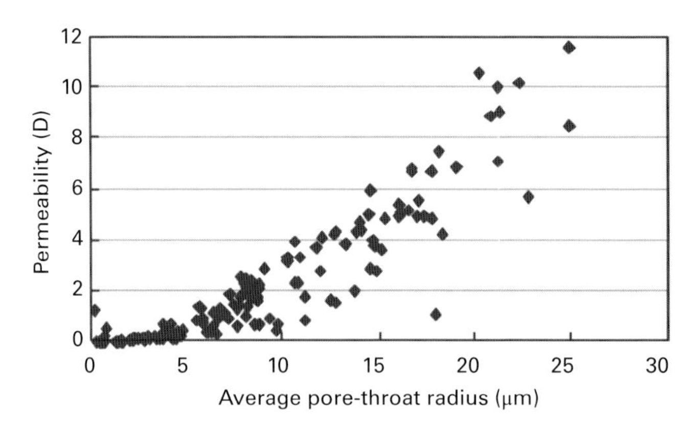

Figure 2-11 Relationship between permeability and average pore-throat radius.

respectively. These figures indicate that the relationship between permeability and maximum pore-throat radius is not very obvious, but the relationship between permeability and average pore-throat radius, permeability, and median radius is obvious. Therefore, the increase of pore-throat median radius or average pore-throat radius will improve permeability on the basis of analyzing the statistical result. When pore-throat radius is larger than 10 μm, the increase of permeability will be faster.

Figures 2-12 and 2-13 show that there is no obvious correlation between permeability and efficiency of mercury withdrawal, permeability, and homogeneity coefficient.

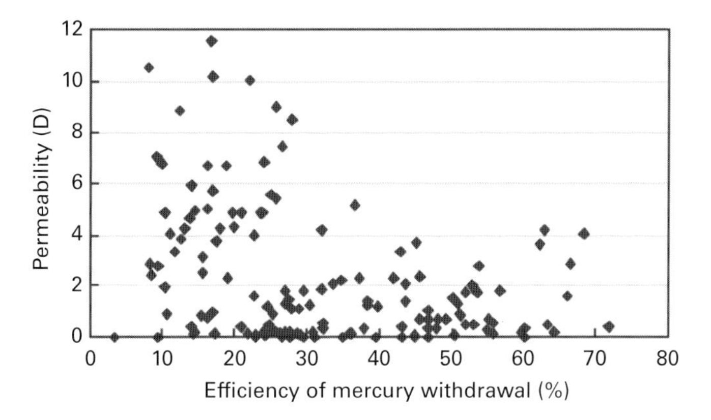

Figure 2-12 Relationship between permeability and efficiency of mercury withdrawal.

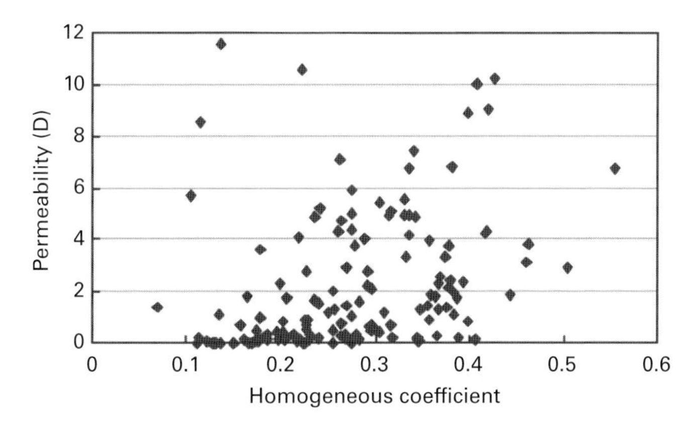

Figure 2-13 Relationship between permeability and homogeneity coefficient.

2.2.2.2 Influence of different sand production rate on pore structure of rock samples

Comparison analysis of particle size of produced sand of wells and samples shows that particle size distribution is almost the same. Therefore, pore size distribution before and after sand production can be measured and compared by reducing formation sand directly to simulate different sand production rate in wellsite.

Experimental methods and processes are as follows:

1. Wash oil and dry sand samples.
2. Fill formation sand into sand tube, weigh, and test original permeability.
3. Evacuate formation sand, saturate it with formation water, and test pore size distribution of samples in sand tube by using conduit flow process (CFP).
4. When the above testing is done, use strong volatile organic solvents to displace formation water in samples to avoid crystallization in rock samples.
5. Dry rock samples again. Use a metal cotton screen to filter and produce 1% and 2% of total sample weight (or produce formation sand directly), evacuate and saturate formation water for samples, and then measure pore size distribution of samples.

Experimental results (Table 2-3, Figure 2-14 to Figure 2-17) and analysis are as follows:

1. Permeability increases by various degrees after sand production. When the sand production rate is 1%, increase of permeability is 12.5% to 41.3% (21.8% average). When the sand production rate is 2%, increase of permeability is 22.6% to 59.8 (41.1% average).
2. The general trend is that the number of coarse throats increases after sand production. When the sand production rate is 2%, the major diameter range of increased throats is 5 to 10 μm, followed by 10 to 15 μm. In extreme situation, the diameter range of increased throats is between 15 and 20 μm or even larger than 20 μm.
3. When sand production rate is 1%, diameter of throats that appears newly is between 2.5 and 5 μm, and individual throats between 5 and 10 μm appear. Diameter of major throats that disappear is between 1 and 2.5 μm, and a few throats whose diameter is between 0.5 and 1.0 μm disappear.
4. For formations with good sorting of initial pore-throat distribution, the original permeability is higher. Sand production rate (less than 2%) has little influence on distribution of pore-throat distribution and permeability.

Table 2-3 Statistics of Influence of Different Sand Production Rates on Pore Structure

					Pore-throat Structure Parameters					
Well No.	*Well Depth (m)*	*Sand Production Rate (%)*	*Permeability (mD)*	*Increment of Perm. (%)*	*Max. Throat Diameter (μm)*	*Change Ratio (%)*	D_{50} *(μm)*	*Change Ratio (%)*	*Major Throat Diameter (μm)*	*Change Ratio (%)*
NB35-2-5	1016.3	0	4792	0	30.38	0	2.18	0	4.07	0
		1.05	5658	18.07	31.28	2.96	3.22	47.71	4.01	–1.47
		2.25	6690	39.61	40.27	32.55	2.74	25.69	7.52	84.77
NB35-2-2	1000.9	0	727	0	12.37	0	2.06	0	1.5	0
		0.83	883	21.46	14.4	16.41	2.25	9.22	3.5	133.33
		2.38	995	36.86	13.03	5.34	2.12	2.91	4.5	200
NB35-2-2	991.6	0	642	0	17.05	0	2.16	0	2.56	0
		1	915	42.52	13.78	–19.18	2.05	–5.09	2.62	2.34
		2	1026	59.81	16.97	–0.47	2.06	–4.63	3.7	44.53
NB35-2-2	991.9	0	973	0	15.88	0	2.12	0	1.88	0
		1.03	1211	24.46	11.29	–28.9	2.04	–3.77	3	59.57
		2.05	1562	60.53	11.91	–25	1.99	–6.13	3.28	74.47
NB35-2-4	1114.1	0	3829	0	23.02	0	2.13	0	2.69	0
		1.09	4534	18.41	18.46	–19.81	1.11	–47.89	3.33	23.79
		2.08	4696	22.64	20.56	–10.69	2.14	0.47	4.01	49.07
NB35-2-4	1116.1	0	3320	0	22.84	0	1.31	0	2	0
		1.01	3851	15.99	22.51	–1.44	1.57	19.85	2.91	45.5
		2	4163	25.39	15.54	–31.96	1.49	13.74	4.08	104

(continued)

Table 2-3 *(Continued)*

Well No.	Well Depth (m)	Sand Production Rate (%)	Permeability (mD)	Increment of Perm. (%)	Pore-throat Structure Parameters					
					Max. Throat Diameter (μm)	Change Ratio (%)	D_{50} (μm)	Change Ratio (%)	Major Throat Diameter (μm)	Change Ratio (%)
NB35-2-5	1001.8	0	5889	0	35.25	0	0.82	0	3.79	0
		1	7157	21.53	36.01	2.16	2.29	179.27	4.08	7.65
		2	7468	26.81	39.88	13.13	1.63	98.78	4.83	27.44
NB35-2-5	1006.8	0	8971	0	35.57	0	2.93	0	4.08	0
		1.02	10,061	12.15	41.11	15.57	1.14	–61.09	4.69	14.95
		2.02	12,009	33.86	42.97	20.8	2.57	–12.29	6.02	47.55
NB35-2-6	1076.9	0	7250	0	35.92	0	2.3	0	4.03	0
		1.01	8409	15.99	38.2	6.35	2.27	–1.3	4.26	5.71
		2.04	9785	34.97	41.59	15.79	2.3	0	5.32	32.01

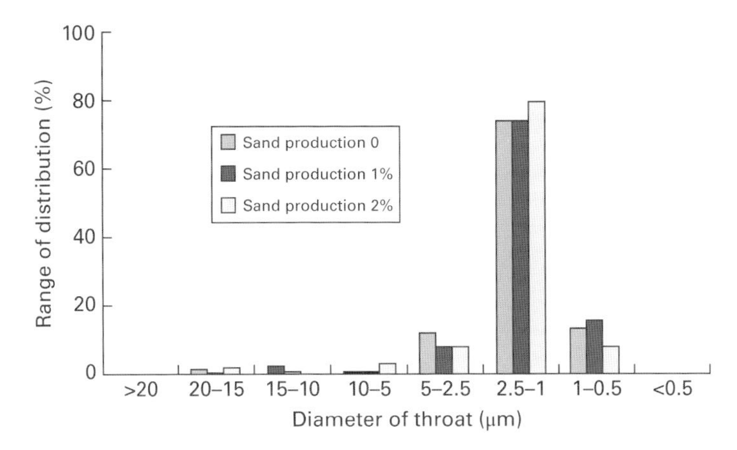

Figure 2-14 Distribution of throats after producing different volumetric sands at interval of 991.60 m of well NB35-2-2.

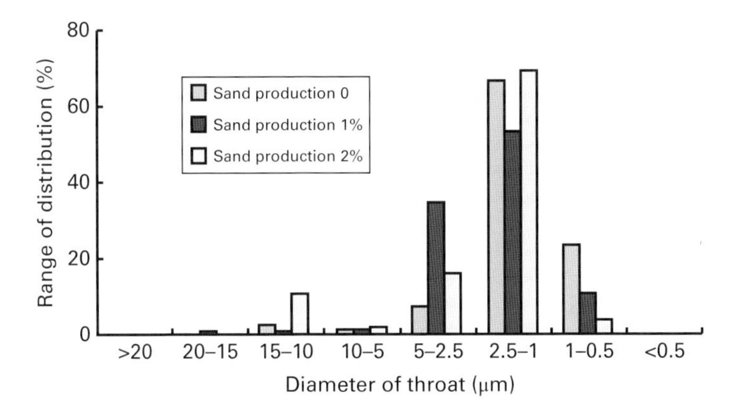

Figure 2-15 Distribution of throats after producing different volumetric sands at interval of 1000.90 m of well NB35-2-2.

5. As for formations with lower permeability, distribution scope of pore-throat is wide and sorting is poor, showing bimodal or multimodal distributions. Figures 2-14 and 2-15 compare pore size distribution of samples at the depth of 991.60 and 1000.90 m of Well NB 35-2-2 before and after producing different amount of sand. Original permeability of samples is less than 800 mD, and distribution scope of effective pore-throat diameter is 1 to 2.5 μm. Permeability increases by 22% to 30% after producing 1% weight of sand sample, and by 37% to 42% after producing 2% weight of sand sample (higher than formations with good

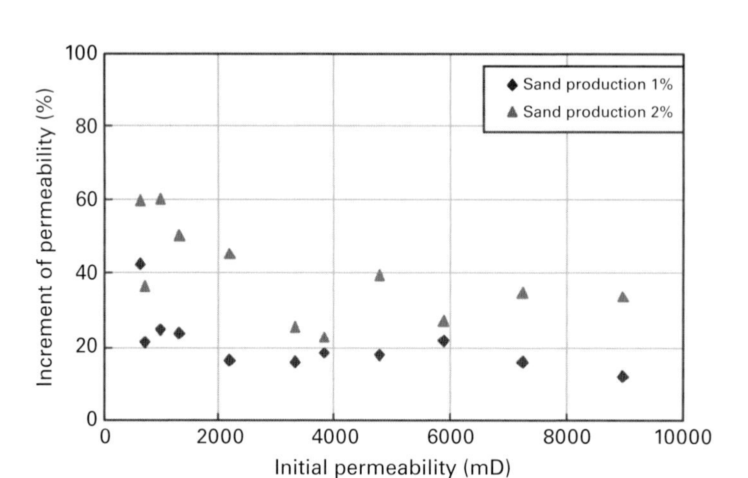

Figure 2-16 Relationship between initial permeability and increment of permeability.

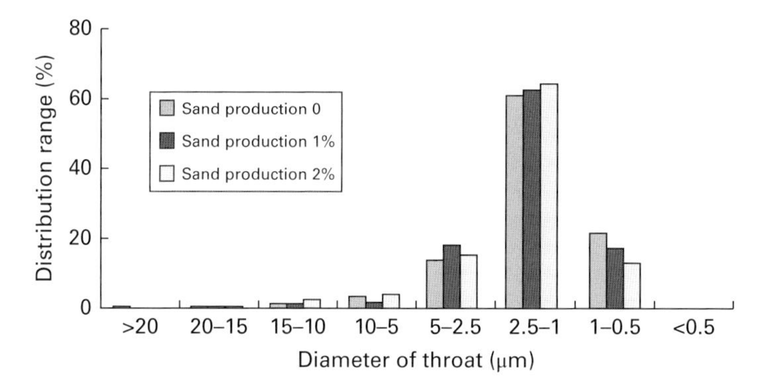

Figure 2-17 Distribution of average throat before and after producing different volumetric sands.

sorting). The result indicates that there are more particles of migration in lower permeability formations and they stay on the surface of throats. In this situation, bridging damage will occur easily and permeability will be reduced. Controlling production drawdown appropriately can lessen the occurrence of bridging damage (Figure 2-16).

6. Comparison of distribution features of pore throat before and after sand shows that the change in peak is obvious. Diameter of effective pore-throat peak of researching area is between 1 and 5 μm (Figure 2-17), and ratio of peak declines.

7. Results of analyzing the effective throats after producing sand by 1% and 2% show that big throats like wormholes will not appear in unconsolidated sandstone. On the basis of statistics of pore structure parameters in Table 2-3, both increase and decrease of maximum diameter of effective pore throat before and after producing sand can be observed, accounting for 50%, respectively. If migrated sand can be produced in the originally coarse throats, the size of the largest throat will increase; on the contrary, it will become smaller. Statistics of D_{50} (diameter of throats when cumulative frequency reaches 50%) indicates that D_{50} tends to increase slightly with the increase of sand production rate, but the increasing ratio is only 8.60%.
8. Comparison of casting SEM of samples shows that the types and sizes of pores do not change significantly before and after sand production. Formation particles in the outlet ends tend to migrate and decrease.

3

Sand Production Rate Prediction and Productivity Index Appraisal

Prediction of sand production rate is a complicated and difficult task because it covers multidisciplinary domains like geomechanics, fluid mechanics, development geology, reservoir engineering, and so on. The process of sand production is affected and controlled by factors such as in situ stress status of formation, components of rocks, features of geomechanics of rocks, fluid properties, processes of drilling and completion, production technologies, and others. Also, the techniques of monitoring sand production status of formation and measuring sand production rate in real time are far from perfect, so there are still a lot of challenges in judging sand production status and verifying theoretical methods of predicting sand production rate.

With the proposal of sand production management, research into building the relationship between sand production rate and well production rate begins to be a focus in this domain. Applications and achievements of limited sand production in unconsolidated sandstone heavy oil reservoirs are being paid increasing attention. Therefore, the solutions for problems such as precise prediction of sand production rate, permeability changes due to continuous sand production, and influence of sand production on well productivity are becoming urgent issues.

This chapter is based on geomechanics and theories of fluid flowing in porous media. Porosity is regarded as a basis parameter of fluid–solid coupling model, which was built from two aspects of flowing of fluid with sand in porous medium and rock skeleton deformation. This model details the changes of permeability after producing sand and the quantitative relationship between sand production rate and well productivity, so the prediction of performance

Sand Production Management for Unconsolidated Sandstone Reservoirs, First Edition.
Shouwei Zhou and Fujie Sun.

of wells producing with limited sand production will be more accurate and realistic, which will help to realize maximum economic profits.

3.1 Evolvement of Sand Production Rate Prediction Method

With the proposals of techniques like CHOPS (Cold Heavy Oil Production with Sand) and sand production management with limited sand production rate, prediction of sand production rate in different production stages of oil and gas wells becomes one of the urgent problems needing to be addressed. After studying published references about sand production around the world, methods of predicting sand production rate on the basis of its history of technical development are summarized as follows: wellsite engineering method, stress–strain model, wormhole model, and fluid–solid coupling model.

3.1.1 Wellsite Engineering Method

In the early stage, the prediction of sand production rate is based on an empirical relationship between sand production rate of oil and gas wells and wellsite parameters such as well depth, acoustic time, well production rate, production drawdown, well productivity, fluid saturation, and so on. Though empirical formulas can also predict sand production rate, to build such relationship requires a lot of wellsite observation data. Therefore, this method cannot be applied in the newly developed oilfields and is constrained by the accuracy of data. In addition, the engineering method lacks a real physical description of sand production process, limiting its application.

In 1999, Mazen Y. Kanj and Younane of Oklahoma University collected a lot of wellsite sand production data and built a neural network model to predict sand production rate. This model is a positive feed structure based on reverse-propagation learning model, whose input parameters include porosity, clay content, geological age and condition, API gravity of oil, and production drawdown. The output data of this model can be used to appraise sand production status of a well, but the neural network model needs to be "trained" first (history matching) and tested effectively before being applied in sand production prediction. The accuracy of prediction and applicable scope depend on the attained sand production data from the wellsite.

In 2001, J. Tronvoll conducted regression processing for a group of sand production data of a pilot well in which sand production was allowed and sand management was applied. The research showed that sand production rate reaches a new peak whenever the production drawdown increases again. Initial sand production rate due to increasing pressure drop is related to actual well drawdown, whereas final sand production rate is related to the total drawdown. Sand production rate during each pressure-fluctuating cycle forms an approximately parabolic relationship with time. The empirical curve indicates the performance of sand production of most wells in which sand production management is implemented, but it does not contain a detailed quantitative description between formation properties and sand production rate.

3.1.2 Stress-Strain Model

In 1994, M. B. Geilikman and colleagues found that the stress around wellbore in formation will change more or less with the depletion of formation energy during continuous oil production. When the pay reaches critical status, sand production occurs. The areas from inside to outside the wellbore can be divided into disturbing zone of sand production, plastic zone, elastic–plastic

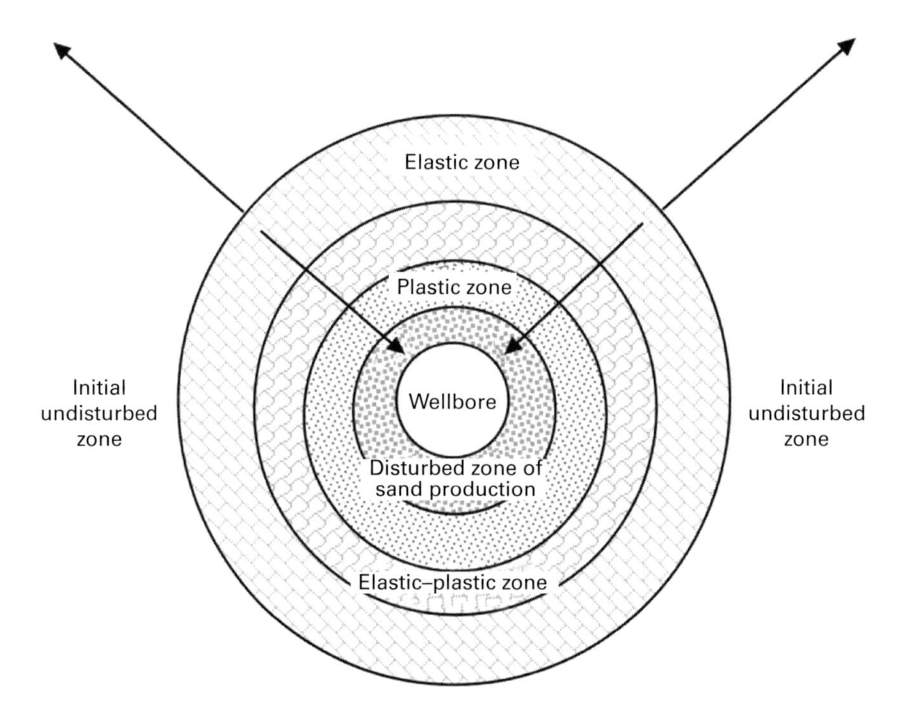

Figure 3-1 Diagram of zones of sand production in formation.

zone, and elastic zone (Figure 3-1). They studied the relationship between sand production rate and features of plastic zone and indicated that sand production mainly results from continuous extension of plastic zone. Rock porosity of plastic zone becomes greater after sand production than that of elastic zone. Physical equation of sand production rate is achieved by using material balance method, showing that porosity changes in plastic zone are resulting from sand production and the porosity change rate is equal to the sand production rate.

$$S_c(t) = \pi h(\phi_p - \phi_e)(R^2 - r_w^2) \tag{3-1}$$

where S_c is cumulative sand production rate (m^3), ϕ_p is porosity of plastic zone (fraction), ϕ_e is porosity of elastic zone (fraction), R is plastic radius (m), and r_w is wellbore radius (m).

This model combines fluid flowing and plastic deformation in porous medium and regards the wellbore radius as an initial value of plastic radium. When the bottom hole pressure of wells drops to the critical status of sand production, particles in rock are changed and sand production occurs. The model proposes physical mechanism of continuous sand production and shows that the boundary of plastic zone will not expand unlimitedly and that the maximum plastic radius is related with initial well production rate and rock strength of formation.

In 1997, Yarlong Wang proposed a method of calculating sand production rate on the basis of analyzing stress–strain distribution around wellbore during production. He believes that well radius keeps constant during sand production, and the sand production rate is equal to total shear deformation around wellbore skeleton. The formula is as follows:

$$S(t,a) = (1-n)r\int_{t_o}^{t}\left[\varepsilon_\theta^e(a,t) + \varepsilon_\theta^p(a,t)\right]\mathrm{d}t \tag{3-2}$$

where S is sand production rate (m^3), n is porosity of rock (fraction), t is sand production time (h), a is wellbore radius (m), ε_θ^e is elastic shear strain (%), and ε_θ^p is plastic shear strain (%).

In 2003, P. J. Van den Hoek and colleagues improved the model proposed by Geilikman and expanded the scope of application of this model. This improved model shows that permeability and porosity in plastic zone will increase after sand production, and formation will go back to its stable state. It means that sand particles tend to form stable sand arching on the surface of screens or nearby perforation tunnels during sand production. Thus, the stress concentration in microstructure of formation will be relieved, and pseudo strength of formation will be improved. The formation, however, is only relatively stable. It will become unstable with greater sand production rate when situations like production drawdown change take place.

In the above-mentioned model, rock is considered as an ideal elastic–plastic material and Mohr-Coulomb yield criteria are used to describe yield the behavior of rock. Also, perforation tunnel is supposed to have spherical or cylindrical cavities with uniform stress, and formation fluid obeys Darcy's law. This model predicts the sand production rate by solving the maximum production rate (or drawdown) and the stable critical radium of sand arching when extension failure of sand arching occurs.

3.1.3 Wormhole Model

In 1996, L. S. K. Fung from CMG and R. C. K. Wong from the University of Calgary derived a three-dimensional model to predict tunnel stability on the basis of geomechanical theory. The end of the tunnel will collapse partly and expand when critical effective stress is reached. The growth direction of the end of tunnel depends on fluid state, fluid properties, geomechanical rock strength, boundary conditions, and geometrical forms of the tunnel (Figure 3-2).

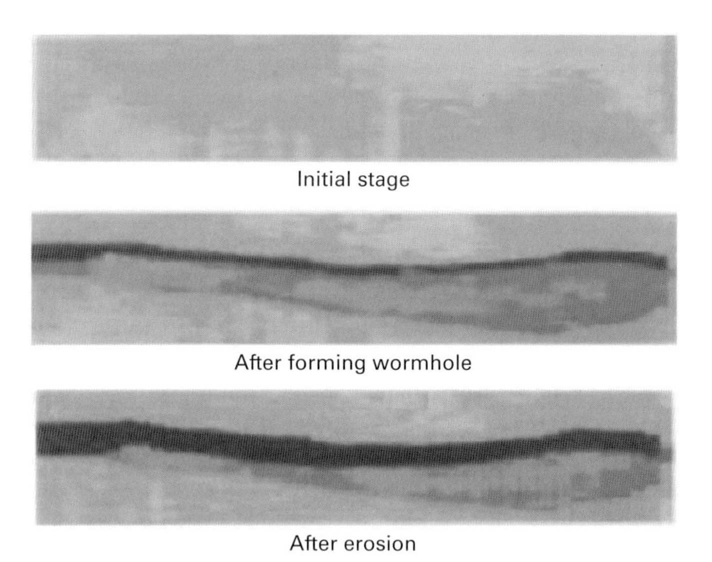

Figure 3-2 Axial computed tomography scan of wormholes (longitudinal section).

In 1998, Bernard Tremblay and colleagues of ARC conducted erosion experiments by using produced sand and oil from the Clearwater group of the Burnt Lake Oilfield of Suncor to simulate the process of CHOPS. The experiments studied the extension of wormhole and distribution of porosity after sand production by using computed tomography scans. The results of experiments are as follows:

1. Wormholes can be formed in unconsolidated sandstone and will expand in high porosity zone.
2. After being formed, wormholes are still filled with sand and their porosity is about 53%. Sand particles in wormholes will be produced continuously after erosion and the porosity of wormhole gets close to 100%.
3. Wormholes will connect two ends of sand-filled model; there is required pressure difference between two ends of the sample and the flowing velocity of the liquid is large enough to enable sand migration.
4. Production rate is relatively low when wormholes are forming and extending, and sand concentration in liquid reaches 44%. After the extension of wormhole is stable, production rate will increase and sand concentration in produced liquid will get lower and lower (within 5%). These results are the same as the actual well performance on the wellsite.

In 1999, Jianyang Yuan and colleagues at ARC built a model to predict wormhole network on the basis of probabilistic active walker (PAW) model. Experiments and theoretical studies show that the major factor determining the formation and growth of wormhole is the ratio between pressure gradient and formation matrix stress, and sand production occurs only when pressure gradient is over a critical value. If the distribution of formation stress is uniform, wormholes will grow in the direction with maximum pressure gradient and along the direction where matrix is the weakest when pressure gradient is the same in all directions. When wormholes extend to a certain length where the pressure difference between the tip of wormhole and formation is very small, the erosion will stop and wormholes will stop extending. Then, the flowing influx begins and sands settled in the wormholes will be carried away. In such situations, the pressure at tips of wormholes will be dropping, and the pressure difference between tips of wormholes and formation is enlarged. The wormholes will grow again when the pressure difference between tips of wormholes and formation reaches critical values. The direction and number of wormholes depend on geomechanics and in situ stress of formations. The wormhole network is similar to the growing mechanism of roots of plants, where there will be more bifurcations when far away from the root (i.e. wellbore).

To simulate the extension of wormholes by using the PAW model is feasible. The potential function is the pressure field of formation, the potential function represents the terrain, and the possibility of pedestrians moving to neighboring points is a function of potential function. Due to changes of adhesive force in different points of formation, the extending direction will change with the extension of wormholes.

One of the common issues of all these models is that it is difficult to determine parameters of geometrical shapes required to simulate the network of wormhole network, and these parameters need to be compared and calibrated with filed data where there are usually multiple solutions. Another issue is that there are different views about the existence of the wormhole network in the formation. Some believed that in the pays where the formation has good cementing strength, collapse may occur instead of wormholes.

3.1.4 Solid–Fluid Coupling Model

The formation is a complex porous medium with solid, liquid, and gas coexisting in it. With the production of oil and gas in fields, the pressure of pores will change and is situ stress will redistribute, resulting in the deformation of skeleton of matrix and changes in flowing conditions of formations, all of which affect fluid flowing in the pore spaces. Therefore, oil and gas production itself is a fluid–solid coupling process.

Based on the above-mentioned principles, many researchers in the domain built various coupling models using different coupling methods and physical assumptions when considering different aspects of sand production.

In 1999, E. Papamichos and colleagues built an integrated framework model for CHOPS. They believed that formation rocks will soften due to stress concentration during heavy oil production, rock strength will decrease, and loose particles will potentially form in formation. Fluid flowing in pores will erode particles of rock, causing the falling of particles from the skeleton of the matrix and the migration of sand. Therefore, the change of porosity can be regarded as a basis parameter of the liquid–solid coupling model: porosity will increase with the erosion and migration of sand particles, and flowing of oil and sand mixture will occur.

The continuity equation of rock matrix is as follows:

$$\frac{\dot{m}}{\rho_s} = \frac{\partial \phi}{\partial t} \tag{3-3}$$

where $\dot{m}$ is erosion rate of skeleton sand per unit volume [kg/(m^3·s)], ρ_s is skeletal particle density (kg/m^3), and ϕ is porosity (fraction).

When rock is in the plastic yielding stage, changes of porosity are very complicated. E. Papamichos believed that the change of porosity results from only sand production, which has obviously underlooked the issue. Actually, the influence of plastic expansion of yielding plastic material on porosity is also very important.

In 2002, Yarlong Wang and colleagues established a fully coupled two-phase model to simulate the flow of heavy oil in porous medium. The model presented continuity equation of two phases of oil and water, fluid motion equation and erosion equation. Volumetric changes in these equations are reflected in an incremental form, meeting the Mohr–Coulomb criterion of elastic–plastic yielding flowing. The change of permeability of formation is determined by either Kozeny–Poiseuille law or Carmen–Kozeny law. The model can be used to calculate increased production rate, cumulative sand production rate, and sand production radius. Compared with the model proposed by E. Papamichos, the difference is that Yarlong Wang considers that in addition to the change of porosity, sand production is also affected by the volumetric strain change ration of rock. The continuity equation of rock matrix is as follows:

$$\frac{\partial \phi}{\partial t} = \frac{\dot{m}}{\rho_s} + (1-\phi)\varepsilon_V \tag{3-4}$$

where ε_V is volumetric change ratio of rock (fraction).

Also, R. G. Wan, J. Wang, and colleagues built a coupled erosion–stress–strain model based on geomechanics of continuous medium. The material balance equation of movable sand (i.e. sand production source) is as follows:

$$\begin{cases} \dot{m}/\rho_s = \lambda(1-\phi) \quad c\|\upsilon_f\| & \|\upsilon_f\| \geqslant \|\upsilon_f^{er}\| \\ \dot{m}/\rho_s = 0 & \|\upsilon_f\| < \|\upsilon_f^{er}\| \end{cases} \tag{3-5}$$

where λ is empirical coefficient (determined by experiments), c is concentration of particles, $\|\upsilon_f\|$ is mode of flowing vector, and $\|\upsilon_f^{er}\|$ is mode of critical flowing vector.

In addition, Dennis Coombe and colleagues from CMG built a hydrodynamic–geomechanic coupling model for CHOPS. The model illustrates mechanical equations in porous elastic and porous plastic media as well as continuity equation of fluid flowing (including movable sand caused by erosion sand), where movable sand is a function of porosity and flowing velocity.

The fluid–solid coupling model is a theoretical model coupling geomechanics and hydrodynamics. The fact that it changes over time is determined by stress field of formations, and its specific form and the corresponding parameters are decided on the basis of experimental data. The set of coupling equations of the above-mentioned model can be solved, or the coupled solution can be achieved by using finite element method to discrete equations of stress equilibrium and the time function of fluid flowing. The rock displacement changing rate and pore pressure changing rate can be solved by using the fully coupling method.

3.2 Sand Production Rate Prediction by Using Fluid–Solid Coupling Model

3.2.1 The Principle of Calculation

Unconsolidated sandstone reservoirs are the typical deformable medium. During the development of oilfields and gasfields, pore pressure of rock around wellbore will drop with fluid flowing into wellbore and the stress around wellbore will redistribute, which will result in changes in effective stress and skeleton of rock. The deformation of rock skeleton changes flowing conditions of formation, which in turn impacts the characteristics of flowing in porous medium. As such, the production process of unconsolidated sandstone reservoirs is coupling flowing (Figure 3-3), where solid phase and liquid phase are mutually included by each other and thus difficult to be distinguished. Unconsolidated sandstone reservoirs can be regarded as continuous media with combination of liquid phase and solid phase, where continuous media of different phases will interact with each other.

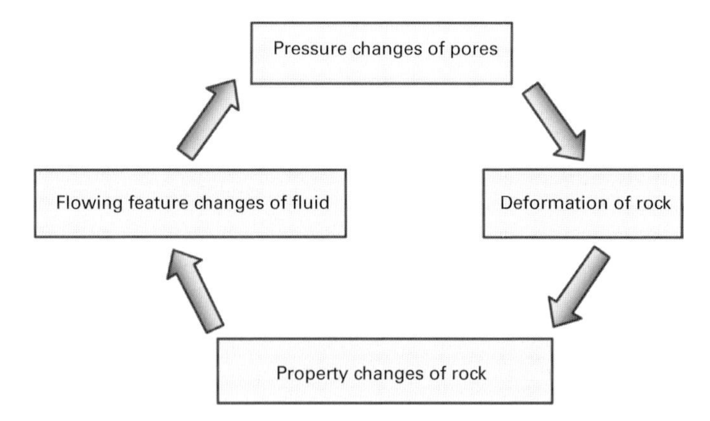

Figure 3-3 Liquid–solid coupling process of sand production in formation.

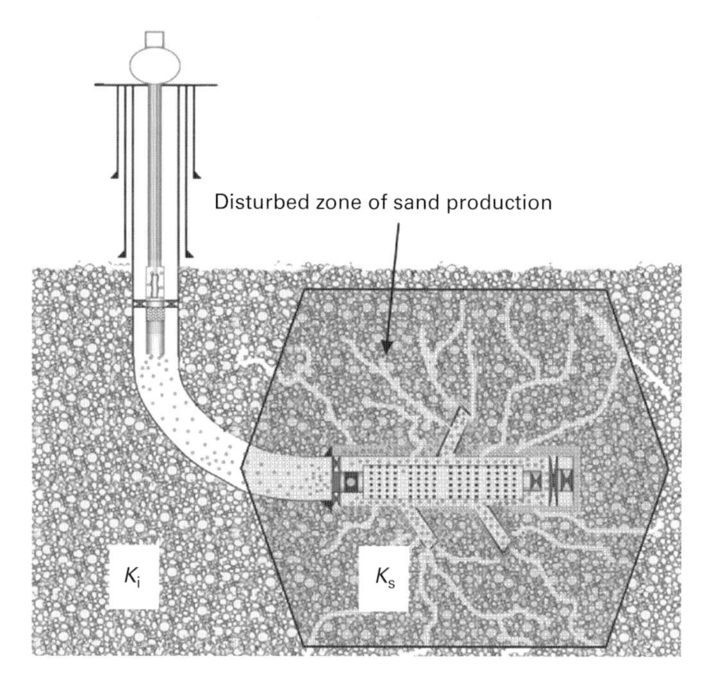

Figure 3-4 Diagram of sand production. K_i is the initial formation permeability, and K_s is the permeability of the sanding zone.

Liquid–solid coupling process of sand production in formation is as follows: at the beginning of production, fluid particles flow in porous media, and sand particles in formation are eroded by fluid flowing. When flowing velocity of fluid reaches a certain value, the filling sand with minimum cohesive force with matrix skeleton will begin to migrate and sand production occurs. With the production of wells, the forces around the wellbore will change, and rock deformation and displacement of rock particle occur. When the production drawdown of wells exceeds critical drawdown, the stress exerted on sand particles is greater than the rock strength and consolidated rock skeleton will become loose in the weak points of rock, destroying rock skeleton. Parts of skeleton sands will be broken and become movable sand eroded by drag force of flowing fluid, and then be carried into wellbore by fluid. After breakdown of skeleton of formation, channels with high permeability will be formed in formation; while movable sand can cause blockage in narrow throats during migration of sand and settling of movable sand. These results will affect hydrodynamics and rock strength of formations in turn, and parameters of rock (e.g. porosity, permeability) will change correspondingly (Figure 3-4).

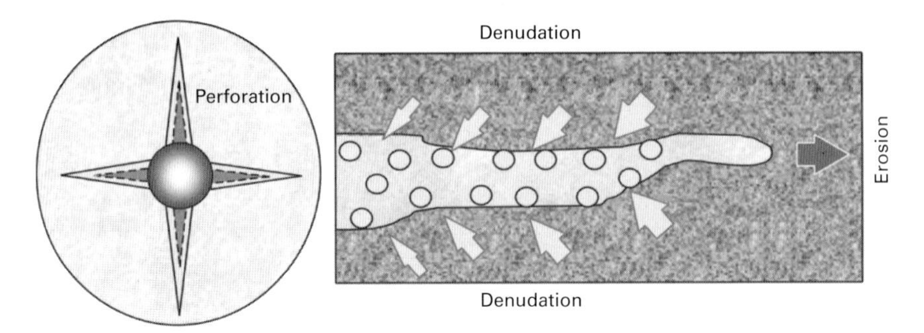

Figure 3-5 Sand production mechanism of erosion model and denudation model.

The model presented in this book to predict volumetric sand production and productivity index appraisal is different from other liquid–solid coupling models. In this model, both erosion sand production in tips of sand production channels and erosion sand production in walls of sand production channels are considered in sand production mechanism (Figure 3-5). The proposed new model is based on liquid–solid coupling flowing principles as well as the improvement of the wormhole model to predict volumetric sand production and well productivity. With respect to sand production channel extension shape, the method of wormhole model is used. In the continuous formations, the model of multiphase flowing is applied to simulate the complex four-phase flowing process (i.e. sand, oil, gas, and water).

In sand production channels near wellbore, transportation of liquid with sand plays an important role in the actual production rate. By building a material conservation relationship between sand liquid transported by channels and liquid with solids flowing to channels through the tips of them, the continuous-type mathematic model can be established to describe the flowing of liquid in channels, in tips of channels, and in continuous formation.

The balancing of in situ stress will be broken when the well is put into production or pressure fluctuation is applied in the formation. For unconsolidated sandstone reservoirs, a plastic zone around the wellbore will be formed, where erosion in the tips of sand production channels and denudation of channel wall occur. Dynamic changes of the plastic zone are illustrated by the radius of plastic zone and physical property model. The extension of tips of channels is described by solid–liquid coupling erosion model, and the dynamic extension is determined by effective stress gradient. When the stress gradient exceeds residual cohesion of sandstone, the sand production channels will grow forward, and the erosion of wall of channels will make channels to extend transversely (e.g. diameter of sand production channels will increase). The erosion model is used to describe this process,

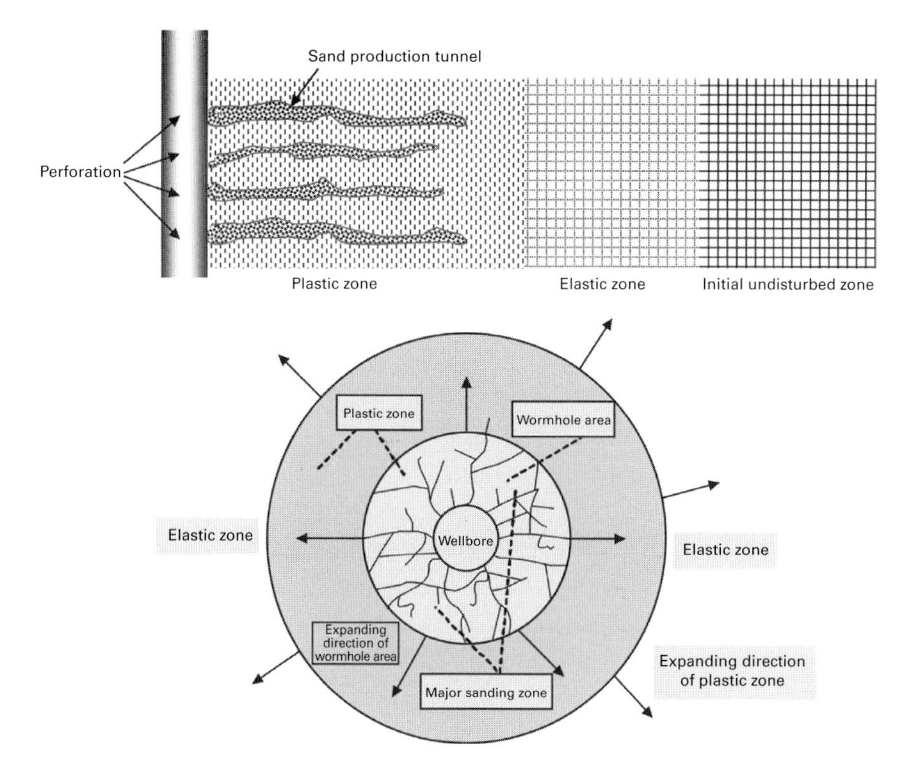

Figure 3-6 Profile and cross section of sand production channels, plastic zone, and elastic zone.

which is mainly controlled by erosion coefficient. Transportation of sand particles in sand production channels and fluid is described by sand liquid transportation model. Figure 3-6 shows the profile as well as a cross section of sand production channels, plastic zone, and elastic zone.

In this model, coupling mechanism involving fluid flowing, stress–strain, and sanding fluid transportation is used to describe sand production process. Production of solid particles is assumed to be the result of disintegration of skeleton, which is due to the loss of cohesive force, expansion of fluid inside rock skeleton, the breakdown of skeleton, and viscous drag forces of fluid.

The viscous drag force of fluid carries solid particles and transports them to channels. The stability of channels depends on quantitative conditions of sanding. On the basis of specific boundary conditions and initial conditions, geometry, constitutive equations, stress equilibrium equation, mass conservation equation, and Darcy law, the problem can be solved.

To simulate the initial status of sand production and transporting process of movable sand particles, models like the seepage model, geomechanics model, and transporting model of fluidized sands are coupled, and their inner material relationship is considered respectively in three different zones, which are the transporting zone of fluidized sands in channels of sanding, extension zone of tips of semispherical channels, and undisturbed zone. The established integrating model helps to simulate volumetric sand production, sanding characteristics over time, and influence of sand production on production rate of wells.

3.2.2 Erosion Model

3.2.2.1 Forming of sanding channels

The key technologies of sand production management are to improve well production rate by forming high permeable channels surrounding wellbore in formations through perturbation under controllable conditions, which are realized by using special methods of completion and development. To achieve this target, the critical condition of forming sanding channels in formation should be measured. Comparing the measured critical condition with production status in fields can help to adjust production operating parameters in fields to achieve the critical condition of forming sanding channels.

B. Tremblay and colleagues at ARC established the critical conditions of forming sanding channels through physical simulation experiments:

$$q_c = \frac{4\pi CKr}{\mu} \tag{3-6}$$

where q_c is critical flowing velocity (cm/s), C is rock strength (atm), K is permeability (D), r is radius of channels (cm), and μ is viscosity (cP).

When the flowing velocity in the tips of channels is greater than critical velocity, sand particles in the tips of channels will be disintegrated and the sanding channels will extend.

3.2.2.2 The forming process of sanding channels

Due to the relatively weak cementation of rock in heavy oil formations, the poor cementing among sand particles is easy to be broken when in situ stress–strain exceeds the critical status. In addition, heavy oil is with high viscosity and better capacity of carrying sand, so sand particles are easily produced with heavy oil and the “wormhole network” will occur gradually surrounding

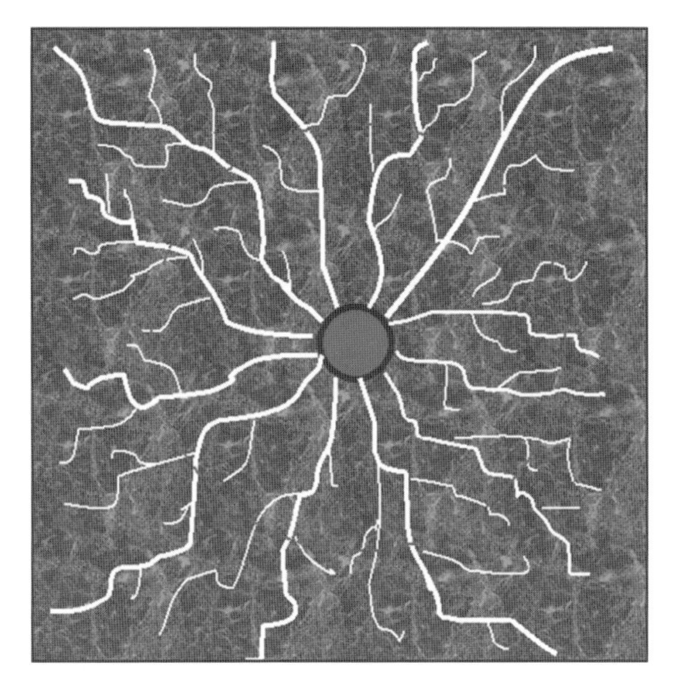

Figure 3-7 Diagram of wormhole network.

the wellbore in formations (Figure 3-7). Therefore, porosity and permeability of formation surrounding wellbore will be improved greatly, increasing the flowing capacity of heavy oil.

Experimental studies show that the key factor influencing the formation and growth of sanding channels is the ratio of rock strength to pressure gradient. When the pressure gradient is greater than critical value, erosion of rock matrix will occur. Under the premise, sanding channels will grow in the direction where the ratio of rock strength to pressure gradient reaches its maximum. Similarly, if pressure gradient in all direction is the same, sand channels will grow toward the direction of soft sandstone matrix.

Branches of sand channels will be formed during the growing of these channels, and the changes of sand channels can be regarded as a self-adjustment process due to external pressure distribution. For instance, when sanding channels are too long and the pressure difference between tips of channels and formation becomes small, the growth of sanding channels will stop, and the erosion will also stop. Then, the denudation function will dominate the sand production process. The denudation process will carry away the movable sand particle deposited in the channels, resulting in enlargement of pressure difference between tips of channels and formation due to the pressure drop in the tip

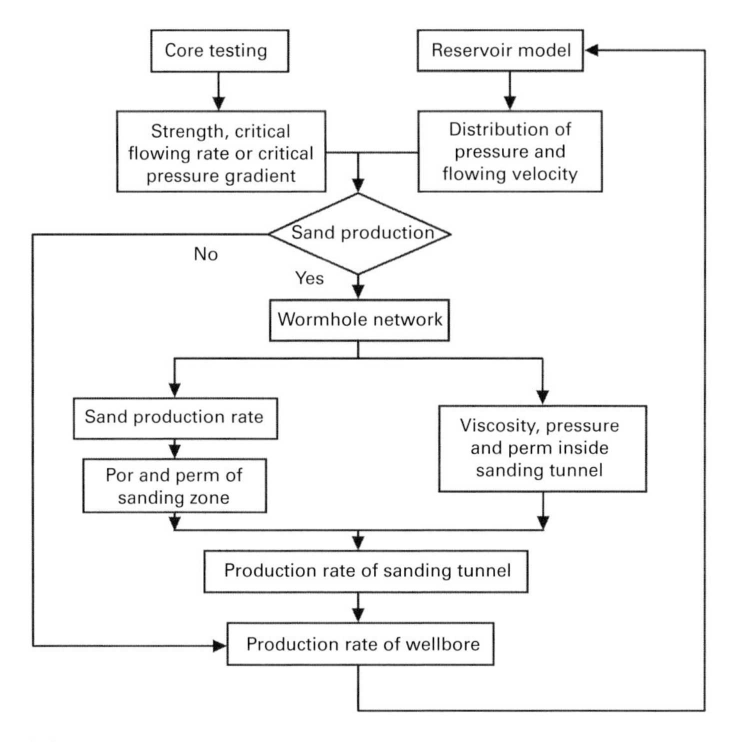

Figure 3-8 Calculation workflow of coupling model of sanding channels.

of channels. When the pressure difference is over critical condition for sanding, the sanding channels will begin to grow again. The pressure distribution makes the growing direction and counts of channels uncertain and the growing point of channels random, so the wormhole network, like the roots of plants, will be formed over time.

The workflow of solving coupling model to simulate the growth of sanding channels, seepage inside reservoirs, and well production is as follows:

3.2.2.3 Boundary of sanding zone

1. Stress distribution of plastic yielding zone
   Assuming only plane deformation occurs when wellbore is drilled, and radial stress σ_r and σ_θ tangential stress surrounding the wellbore are the maximum values, the stress distribution around the wellbore in plastic yielding zone can be achieved according to elastic and plastic mechanics of rock:

$$\sigma_{r_p} = \left(p_w + \frac{2S_0N}{N^2-1}\right)\left(\frac{r}{r_w}\right)^{N^2-1} + \left(1-N^2\right)r^{N^2-1}\alpha\int_{r_w}^{r} pr^{-N^2}\,dr + \frac{2S_0N}{1-N^2} \quad (3\text{-}7)$$

$$\sigma_{\theta_p} = N^2\left(p_w + \frac{2S_0N}{N^2-1}\right)\left(\frac{r}{r_w}\right)^{N^2-1} + N^2\left(1-N^2\right)r^{N^2-1}\alpha\int_{r_w}^{r} pr^{-N^2}\mathrm{d}r$$
$$+\left(1-N^2\right)\alpha p + \frac{2S_0N}{1-N^2} \tag{3-8}$$

where

$$N = \tan\left(45° + \frac{\varphi}{2}\right) \tag{3-9}$$

where σ_{r_p} is radial stress in plastic yielding zone (MPa), σ_{θ_p} is tangential stress in plastic yielding zone (MPa), p_w is stress surrounding wellbore (MPa), r_w is wellbore radius (m), S_0 is cohesive force (MPa), φ is internal friction angle (°), α, Biot coefficient, and p is pore pressure (MPa).

2. Stress distribution in elastic zone
Assuming that the elastic–plastic interface is r_c, the radial stress is σ_{r_c}, and radial stress at the outer boundary of elastic zone $r = r_0$ is σ_{r_0}, then stress distribution in the elastic zone is as follows:

$$\sigma_{r_e} = \sigma_{r_c} + \left(\sigma_{r_0} - \sigma_{r_c}\right)\frac{1-(r_c/r)^2}{1-(r_c/r_0)^2} + \frac{1-2\mu}{2(1-\mu)}\frac{\alpha}{r^2}\int_{r_c}^{r} pr\mathrm{d}r$$
$$-\frac{1-2\mu}{2(1-\mu)}\frac{1-(r_c/r)^2}{1-(r_c/r_0)^2}\frac{\alpha}{r_0^2}\int_{r_c}^{r_0} pr\mathrm{d}r \tag{3-10}$$

$$\sigma_{\theta_e} = \sigma_{r_c} + \left(\sigma_{r_0} - \sigma_{r_c}\right)\frac{1-(r_c/r)^2}{1-(r_c/r_0)^2} + \frac{1-2\mu}{2(1-\mu)}\frac{\alpha}{r^2}\left(\int_{r_c}^{r} pr\mathrm{d}r - pr^2\right)$$
$$-\frac{1-2\mu}{2(1-\mu)}\frac{1+(r_c/r)^2}{1-(r_c/r_0)^2}\frac{\alpha}{r_0^2}\int_{r_c}^{r_0} pr\mathrm{d}r \tag{3-11}$$

where σ_{r_e} is radial stress in elastic zone (MPa), σ_{θ_e} is tangential stress in elastic zone (MPa), and μ is Poisson's ratio of rocks.

As stresses in the boundary of elastic zone and plastic zone ($r = r_c$) are continuous, the continuous conditions are applied for the stresses in elastic zone and plastic yielding zone:

$$\begin{cases} \sigma_{r_p}\Big|_{r=r_c} = \sigma_{r_e}\Big|_{r=r_c} \\ \sigma_{\theta_p}\Big|_{r=r_c} = \sigma_{\theta_e}\Big|_{r=r_c} \end{cases} \tag{3-12}$$

Subscripts e and p denote elastic solution and plastic solution, and radius of boundary of elastic–plastic zone can be achieved by solving Equation (3-12).

3.2.2.4 Extension principles of sanding zone

With the development of fields, the ranges of pressure drop will enlarge continuously and plastic yielding zone will extend, but the extending velocity will slow down gradually. Assume that there exists exponential relation between plastic yielding radius and producing time, e.g.:

$$r_c = r_c(t) = r_{c0}\left(1 + r_c^*(t)\right) \tag{3-13}$$

There,

$$r_c^*(t) = \int_0^t \frac{\alpha}{\sqrt{2\pi}\sigma(t+1)} e^{-\frac{[ln(t+1)-\beta]^2}{2\cdot\sigma}} dt \tag{3-14}$$

where r_{c0} is initial radius of plastic yielding zone, t is production time, and α, β, σ are logarithm normal distribution coefficient (Figure 3-9).

Actually, the extending velocity of the radius of plastic yielding zone is the function of both production time and well drawdown, e.g. $r_c = r_c(t, \Delta p)$. Radius

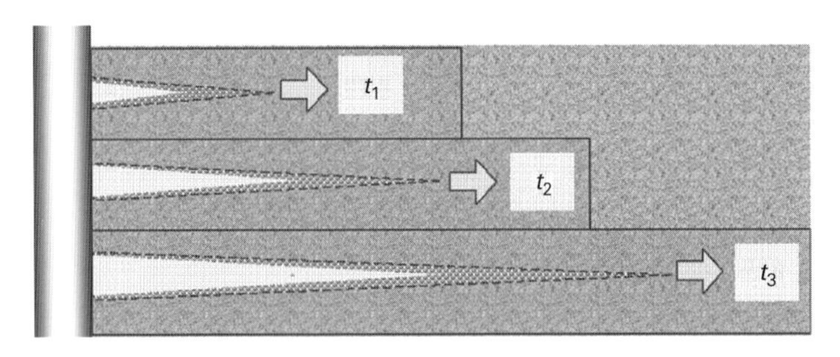

Figure 3-9 Extending diagram of sanding zone over time ($t_1 < t_2 < t_3$).

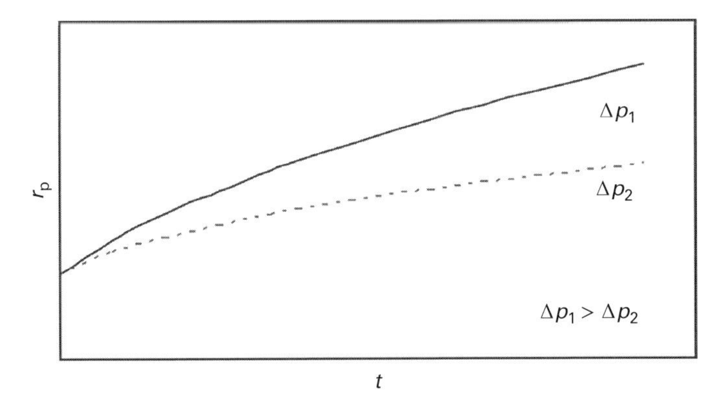

Figure 3-10 Relationship between radius of plastic yielding zone and production time/well drawdown.

of plastic yielding zone extends faster under the condition of larger well drawdown. The relationship between radius of plastic yielding zone and well drawdown, production time is shown in Figure 3-10.

3.2.2.5 Distributing features of sanding channels

The forming process of sanding channels and extending principles of sanding zones determine that the sanding channels are with fractal characteristics. With the increase of sanding branches, the radius of channels will become smaller. For ease of analysis, it is assumed that sand production occurs in plastic yielding zone, and the sanding channels formed in sanding zones have the same influence on parameters of formation properties.

The counts of sanding channels $N(r)$ whose distance is r away from wellbore can be calculated by the following function:

$$N(r) = N(r_0)\left(\frac{r}{r_0}\right)^{d-1} \tag{3-15}$$

where r_0 is radius of wellbore (m), $N(r_0)$ is counts of channels around the wall of wellbore, r is radius of a certain point in the formation (m), and d is irregularity factor of sand production zone.

Assuming the reservoir pressure and bottom hole pressure are remaining constant, the pressure at the tips of channels will increase when sand production channels are growing, leading to the decrease of pressure differential between reservoir and the end of channels, and then the reduction of channel diameter and extending speed at the tip of channels. Therefore, the average diameter is a decreasing function of distance from wellbore.

$$D(r) = D(r_0)\left(\frac{r}{r_0}\right)^{-\varepsilon} \tag{3-16}$$

where ε is a small positive number.

3.2.2.6 Distribution of permeability in formation

The absolute permeability in sanding channels can by calculated using the correlation of Kozeny–Carmen:

$$K = 5.629 \times 10^{-3} \frac{d_m^2 \phi_s^3}{(1-\phi_s)^2} \tag{3-17}$$

where d_m is average diameter of sand particles (μm), ϕ_s is porosity of sand production channels (fraction), and K is absolute permeability (μm^2).

The effective permeability of oil phase can be expressed as:

$$K_c = KK_{ro} \tag{3-18}$$

where K_{ro} is relative permeability of oil phase.

Alternatively, the empirical correlation can be used:

$$K = K_0 \frac{\phi^3}{(1-\phi)^2} \tag{3-19}$$

where K_0 is initial permeability (μm^2).

According to particle size distribution and sieve analysis curve, the correlation achieved from experimental data of sand production with porosity and permeability, the improvement potential of formation properties due to sand production can be estimated. One set of the experimental data are as follows:

1. When the average concentration of produced sand is 0.50% by weight, the average increasing ratio of permeability is 12.0%.
2. When the average concentration of produced sand is 2.63% by weight, the average increasing ratio of permeability is 73.94%.
3. When the average concentration of produced sand is 8.14% by weight, the average increasing ratio of permeability is 501.63%.

For a circular reservoir, the pressures of internal and external boundaries are constant, and fluid parameters and geomechanics parameters along the wellbore are distributed radically and symmetrically. The drainage area is divided into elastic deformation zone and plastic deformation zone. The parameters of fluid and rock might be different from one zone to the other zone, but they will be the same if in the same area. The zone away from wellbore is slightly affected by the pressure drop, and elastic deformation that occurred can be restored under certain conditions, so permeability in the elastic area can be regarded as initial permeability. Assuming the permeability in plastic deformation area is K_c and permeability in elastic area is K_e, the average permeability of formation can be calculated using seepage mechanics:

$$\bar{K} = \frac{\ln\left(r_e / r_w\right)}{\left(1 / K_e\right) ln\left(r_e / r_c\right) + \left(1 / K_c\right) ln\left(r_c / r_w\right)} \tag{3-20}$$

Sand production not only changes formation permeability but also reduces the strength of reservoir rock. The empirical correlation between internal friction angle and cohesive force with porosity is written as follows:

$$\varphi = \varphi_0 \frac{1-\phi}{1-\phi_0} \quad C = C_0 \frac{1-\phi}{1-\phi_0} \tag{3-21}$$

3.2.3 Denudation Model

3.2.3.1 Equation of denudation

The movable sand particles in formation will be carried to the wellbore continuously by formation fluid when producing with sand. The disturbing zone of sand production initiates from the area surrounding wellbore and gradually extends to the internal area of reservoir. The conventional geomechanics equation can only be used to describe the plastic deformation outside the disturbing zone of sand production. The improved formula proposed by Vardoulakis and colleagues can be applied in the area between denudation face and disturbing zone of sand production.

Continuity equation of solid skeleton is as follows:

$$\frac{\dot{m}}{\rho_s} + \left(1-\phi\right)\frac{\partial \varepsilon_v}{\partial t} = \frac{\partial \phi}{\partial t} \tag{3-22}$$

Continuity equation of fluidized sand is as follows:

$$\frac{\dot{m}}{\rho_s} + \frac{\partial(c\phi)}{\partial t} = \nabla \cdot (cq_i) \tag{3-23}$$

Denudation equation is as follows:

$$\frac{\dot{m}}{\rho_s} + \lambda(1-\phi)c\sqrt{q_i \cdot q_i} \tag{3-24}$$

where c is percentile of movable sands over total sands and fluid by volume (e.g. sand concentration), q_i is total volume of sands and fluid through an unit area per unit time [m^3/(m^2·s)], ρ_s is density of skeleton particles, and λ is denudation coefficient, describing the damage process of formation related with denudation, L-1.

In 1998, Papamichos proposed the empirical correlation between denudation coefficient and rock material/plastic volumetric strain:

$$\lambda = \begin{cases} 0 & \Delta\varepsilon^{p} \leqslant \Delta\varepsilon_c^{p} \\ \lambda_1\left(\Delta\varepsilon^{p} - \Delta\varepsilon_c^{p}\right) & \Delta\varepsilon_c^{p} \leqslant \Delta\varepsilon^{p} \leqslant \Delta\varepsilon_c^{p} + \lambda_2/\lambda_1 \\ \lambda_2 & \Delta\varepsilon_c^{p} + \lambda_2/\lambda_1 \leqslant \Delta\varepsilon^{p} \end{cases} \tag{3-25}$$

where λ_1, λ_2 are experimental testing constants, $\Delta\varepsilon^{p}$ is plastic volumetric strain, and $\Delta\varepsilon_c^{p}$ is critical value of plastic volumetric strain during initial sand production.

3.2.3.2 Transportation model of fluidized sands in tunnels

Assuming the channels of sand production initiate from the perforated tunnels, the channels will shortly extend to its maximum cross-sectional area. Supposing that the diameter of channels is constant, the channels will only extend forward from the tips of sanding channels due to the less effective radial stress gradient and greater compressibility at the walls of channels after forming sand production channels. At the tips of sand production channels, the pressure gradient will increase gradually due to the convergence of three-dimensional flowing and the occurrence of gas bubbles. In the plastic zone of unconsolidated reservoirs, only small residual stress exists in rock, so sand production can occur in any suitable area in the plastic zone and will not happen outside it. The sand production area in plastic zone can be identified by comparing effective radial stress and residual cohesive force resulting from residual cementation and capillary pressure.

The sanding channels can only expand forward when the pressure at the tips of sanding channels is lower than critical pressure of sand production. On the basis of linear elastic stress analysis, the maximum shear stress should occur at the walls surrounding channels, so the seepage pressure surrounding cavity wall determines the time of the occurrence and the production rate of sanding.

When the pressure at the tips of sanding channels is exceeding a certain value, where the pressure gradient is less than the critical pressure gradient of sanding, the extension of wormholes will stop. According to the principle of mass conservation, the quality of the inflow and outflow within a cross-sectional area A of sanding channels is equal to the increased volumetric of sanding channels.

$$\upsilon_s A - Q_o = A\dot{\gamma}(t)(\phi_s - \phi_m) \tag{3-26}$$

where

$$Q_o = \phi_s A \upsilon_s \tag{3-27}$$

where $\dot{\gamma}(t)$ is function of sanding radium over time, Q_o is oil rate, v_s is flowing velocity of sand particles, ϕ_s is porosity of sanding tunnels (fraction), and ϕ_m is porosity of plastic zone (fraction).

Introducing Equations (3-26) and (3-27), the mass conservation equation can be simplified to the following:

$$(1-\phi_m)\dot{\gamma}(t) = (1-\phi_s)\left[\dot{\gamma}(t) - \upsilon_s\right] \tag{3-28}$$

According to the above-mentioned calculation principle, the denudated sand volume flowing into each sanding tunnel is the same when denudation occurs at the tips of sanding tunnels, and it forms a linear relationship with the extending distance. Therefore, massive sand production will happen in the early stage of development and results in high sand concentration at the early stage. With the increasing rate of oil flowing through walls of sanding tunnels, sand concentration will reduce and sand production will stop eventually. Also, the volume of sand production at the tips of tunnels in formation and the cumulative sands produced in the wellbore can be calculated.

3.2.3.3 Distributing model of flowing pressure and velocity in sanding tunnels

There is a pressure drop along the path from tips of sanding tunnels to wellbore due to the energy loss caused by the friction with walls of sanding tunnels and high viscosity fluid with sands. On the basis of geomechanical properties of reservoirs, hydrodynamic characteristics and denudated features of fluid mixed with sand particles, the extending velocity of sanding tunnels can be achieved.

Assuming that the sanding tunnels are cylindrical, the following equation can be achieved according to mass conservation of solids:

$$\frac{\partial}{\partial x}\left[\rho_b\left(1-\phi_s\right)\upsilon_s\right]=\frac{\partial\left[\rho_b\left(1-\phi_s\right)\right]}{\partial t}+R\left(x,t\right) \tag{3-29}$$

where ρ_b is density of rock skeleton (g/cm^3), and $R(x)$ is flowing rate through walls of sanding tunnels.

The transportation of fluidized sands in tunnels can be simplified to one-dimension Darcy seepage:

$$\frac{\partial}{\partial x}\left(\rho_s\frac{K_s\partial p_s}{\mu_s\partial x}\right)=\frac{\partial\rho_s}{\partial t}+R\left(x\right) \tag{3-30}$$

where K_s is permeability in sanding tunnels (μm^2), ρ_s is average density of fluidized sands (g/cm^3), and μ_s is viscosity of fluidized sands (mPa·s).

$$r_w<x<\dot{\gamma}\left(t\right)$$

R(x) represents the flowing rate through walls of tunnels as the source term and is defined as follows under steady state conditions:

$$R\left(x\right)=-\frac{2\pi r_c k_m}{\mu_m\ln\left(r_d/r_c\right)}\left(p_s\left(x\right)-p_d\right) \tag{3-31}$$

Subscripts *c* and *d* here represent the sanding cavities and drainage boundary. The relationship between average density of fluidized sands and porosity of sanding tunnels is as follows:

$$\rho_s=\rho_b\left(1-\phi_s\right)+\rho_f\phi_s \tag{3-32}$$

where ρ_f is fluid density (g/cm^3).

If ignoring the compressibility of mixture of sands and fluid, the pressure distribution of the mixture along the sanding tunnels can be expressed as:

$$p_s\left(x\right)=p_w+\Theta\int_{k_s}^{x}\frac{\mu_s}{\left(1-\phi_s\right)\rho_s K_s}\left[\int^{\zeta}R\left(\xi\right)d\xi\right]\mathrm{d}\zeta \tag{3-33}$$

where

$$\Theta=\left(p_m\left(t\right)-p_w\right)\left[\int_{r_w}^{\dot{\gamma}(t)}\left[\frac{\mu_s}{x\left(1-\phi_s\right)\rho_s K_s}\int^{x}R\left(\xi\right)\mathrm{d}\xi\right]\mathrm{d}\zeta\right]^{-1} \tag{3-34}$$

Porosity of sanding tunnels ϕ_s is a function of x, and it varies continuously between the wellbore and the tips of tunnels. Parameters p_w and p_m represent the pressure at the wellbore and the tips of sanding tunnels. Thus, the boundary condition at the wellbore and the tips of tunnels can be expressed as:

$$p_s(r_w, t) = p_w \tag{3-35}$$

$$p_s(\dot{\gamma}(t), t) = p_m(\dot{\gamma}(t), t) \tag{3-36}$$

$$\frac{K_s}{\mu_s}\frac{\partial p_s(\dot{\gamma}(t), t)}{\partial x} = (1-\phi_m)\frac{\partial u(\dot{\gamma}(t), t)}{\partial t} \tag{3-37}$$

$$\phi_s \frac{K_s}{\mu_s}\frac{\partial p_s(\dot{\gamma}(t), t)}{\partial x} = \frac{K_m}{\mu_m}\frac{\partial p_m(\dot{\gamma}(t), t)}{\partial r} \tag{3-38}$$

where u is displacement of solid phase particles of reservoir.

Assuming the mobility and porosity of the mixture of sands and fluids are constant, the pressure distribution along the tunnels can be approximately expressed as:

$$p(x,t) = \frac{p_w \sinh\kappa\left[\dot{\gamma}(t) - x\right] + p_m(t)\sinh\kappa\left[x - r_w\right]}{\sinh\kappa\left[\dot{\gamma}(t) - r_w\right]} \tag{3-39}$$

3.3 Productivity Appraisal Methods of Wells Producing with Sands

When producing with sands, large cavity will be formed in some permeable areas surrounding the wellbore due to the effect of denudation, as a result of which the properties of formation and permeable conditions will be improved and well productivity will be enhanced. Currently, there are three theoretical models to simulate this mechanism.

3.3.1 Model of Skin Factor

According to the principles of equivalent radius of borehole, the improved flowing condition due to sanding denudation can be quantified as a "negative" skin to improve well productivity index in the model (Figure 3-11). The advantage of this model is that it is unnecessary to describe complicated

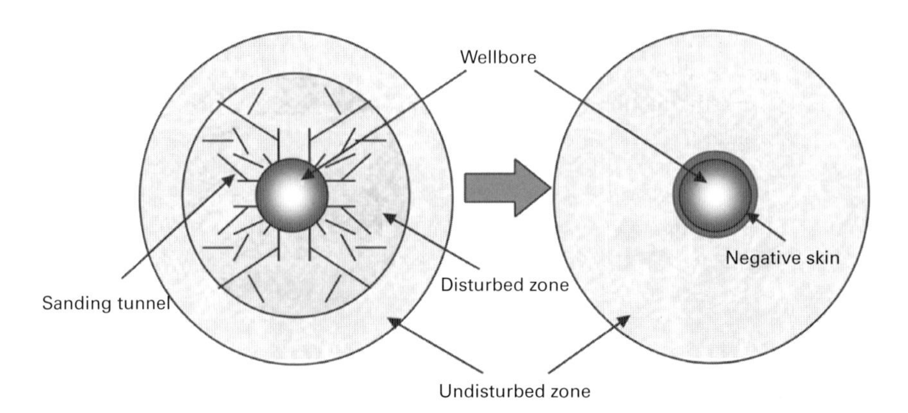

Figure 3-11 Illustration of skin factor model.

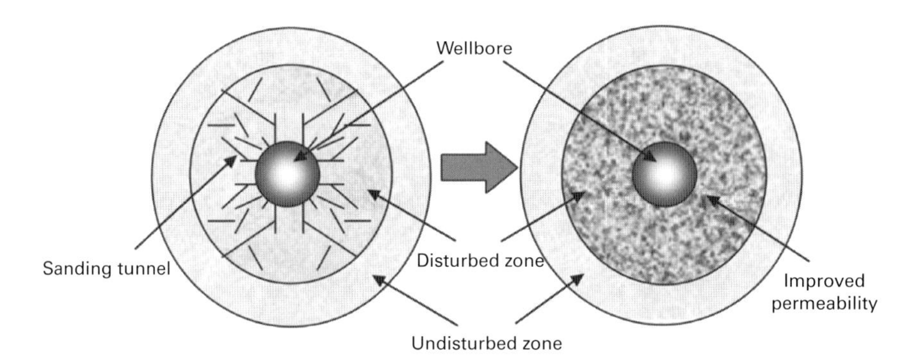

Figure 3-12 Illustration of improved permeability model.

dynamics within the reservoir, requiring fewer parameters and fast computation. The disadvantage is that the increase of production rate due to sand production is only related with wellbore and it cannot simulate the permeable improvement within reservoirs surrounding wellbore during field development. This model can be used for the productivity appraisal at the early stage and for comparison of different well completions.

3.3.2 Improved Permeability Model

The model is to estimate the permeability improvement due to changes in porosity surrounding wellbore based on the relationship between porosity and permeability when producing with sands (Figure 3-12). The advantage of this model is that it reflects the changes of permeable conditions resulting from the sanding channels, and the drawback is that it cannot simulate flowing characteristic inside the sanding tunnels, leading to a larger deviation from real productivity.

3.3.3 Equivalent Wellbore Model of Sanding Tunnels

This model predicts the expanding dynamics of sanding channels by building fractal network model of sanding tunnels, and combining with formation conditions and well completion, including the number and size of sanding channels. This model helps calculate the flowing inside each sanding channel and estimate its contribution to real production rate by considering the sanding tunnels as additional producing wells according to the seepage conditions inside sanding channels and external conditions of reservoirs (Figure 3-13). The model is the most perfect one for appraising productivity when producing with limited sand production, which has been applied and validated in a large number of practical applications like history matching and dynamic prediction of oil wells producing with sands.

In this model, the disturbing zone of sand production is regarded as a virtual well, which does not exist physically; the size of wellbore is determined by the hollowed volume due to denudation in the disturbing zone; and the interval of virtual well is determined by the range of sanding zone. This model is used to calculate the influence of disturbing zone of sand production on well productivity on the basis of pressure inside the sanding channels, reservoir pressure, and mobility inside and outside the tunnels.

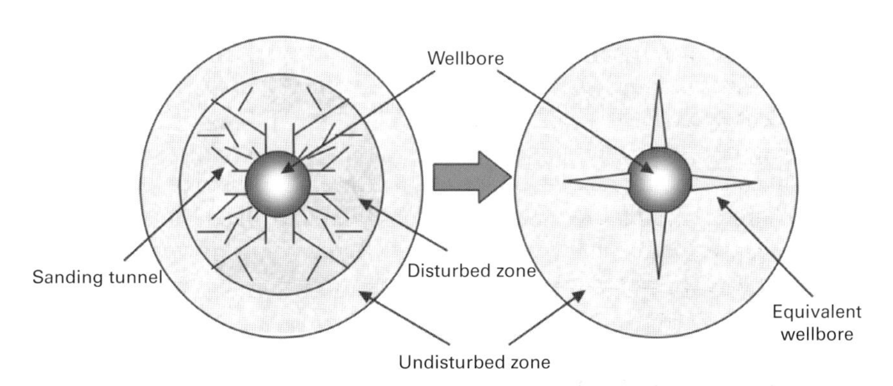

Figure 3-13 Diagram of equivalent wellbore model of sanding tunnels.

3.3.4 Solving Process

Compared with conventional production wells, one outstanding feature of predicting the productivity of sanding wells is that oil is produced from not only the intervals of well but also the tips and walls of sanding tunnels. The complexity of this model lies in the need of predicting not only the dynamic extension of sanding tunnels but also the change between sanding channels and reservoir.

The expansion of sanding channels is a dynamic process and, therefore, productivity predition of sanding wells is also a dynamic process (Figure 3-14). The steps of predicting volume of sand production and productivity are as follows:

1. Obtain formation parameters like rock strength, internal friction angle using geomechanical experiments, or other methods.
2. Evaluate the trend of rock strength changing with formation effective stress, fluid saturation, etc.
3. Estimate the critical pressure and velocity of sand production in formation.
4. Measure the denudation coefficient of rock using flooding and sand production experiments.
5. Set the initial time for simulation.
6. Calculate the distribution of reservoir pressure, fluid saturation, and flowing velocity using numerical simulation software.
7. Determine whether sand production will occur and the radius of sanding area on the basis of the failure criteria of rock and stress-strain.
8. Calculate the expansion of sanding zone over time using the erosion model.
9. Calculate the number and size of sanding tunnels within sanding zone using fractal geometrical model.
10. Calculate the flow rate of fluid/sand at the tips of sanding channels according to the pressure of sanding tunnels and reservoir.
11. Calculate the flow rate of fluid/sand through the walls of sanding tunnels using erosion model.
12. Calculate the distribution of porosity, permeability, viscosity of mixture, and the concentration of sand/fluid within sanding tunnels.
13. Calculate the distribution of pressure within sanding tunnels.
14. Calculate the rate of sand/fluid using the pressure differential between sanding channels and reservoir.
15. Obtain the production rate of wells producing with sands by adding the rate of sanding channels to wellbore.
16. Increase the time step.
17. Repeat steps (6) to (16).

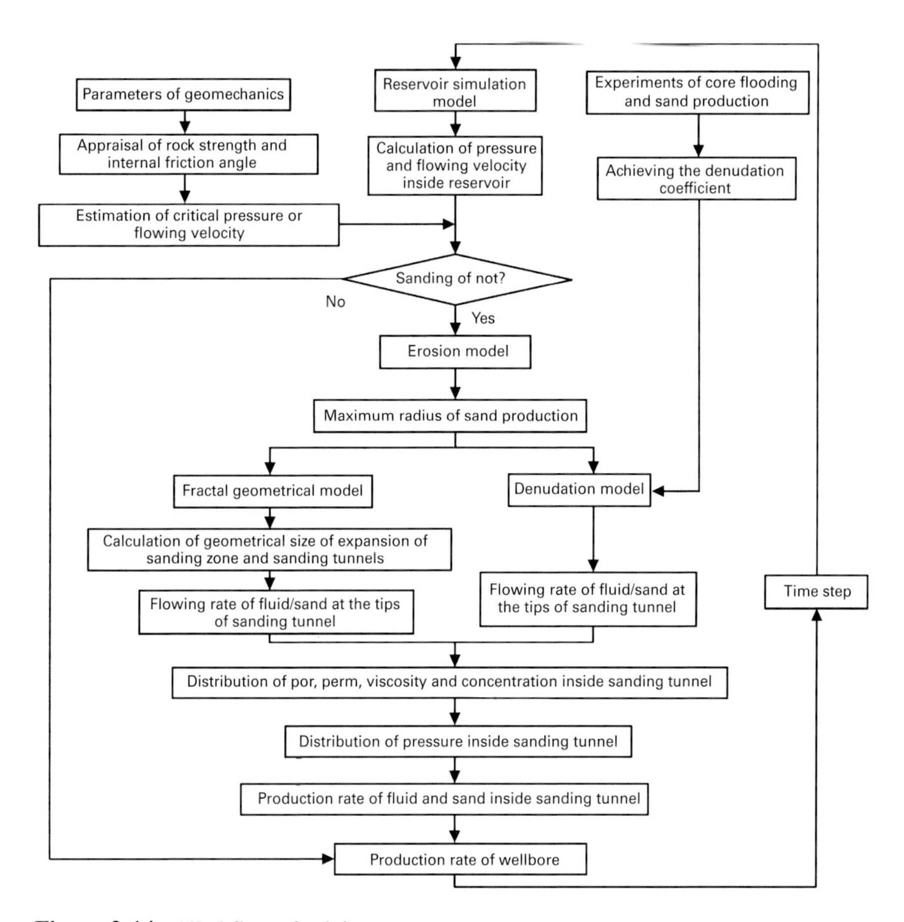

Figure 3-14 Workflow of solving process.

3.4 Cases of Increasing Production Rate by Allowing Sand Production

On the basis of the above solid–fluid coupling model, the software of predicting sand production is developed and applied in an unconsolidated sandstone reservoir, which allows limited sand production to achieve the sanding prediction, sanding features, dynamic extension of sanding disturbing zone, and the relationship between production rate and the volume of produced sands.

3.4.1 Required Parameters for Calculation

The required parameters for calculation in this model are shown in Table 3-1.

Table 3-1 Basic Parameters Used for Calculation

No.	*Parameter*	*Data*
1	Initial reservoir pressure (MPa)	12.5
2	Thickness (m)	10.0
3	Drainage radius (m)	200.0
4	Horizontal permeability (D)	2
5	Vertical permeability (D)	0.5
6	Effective porosity	0.25
7	Rock cohesion force (MPa)	0.85
8	Denudation coefficient (m-1)	12.0
9	Porosity of wormhole	0.78
10	Viscosity (mPa·s)	500

Table 3-2 Drawdown Control Point

Data points	*Time (h)*	*Pressure (MPa)*
a	0	12
b	2	11
c	18	4

3.4.2 Controlling the Bottom Hole Pressure

Reducing the bottom hole pressure continuously in the 18 hours after putting on production to helps form and expand the sanding disturbing zone. Then try to keep the drawdown constant to control the sanding area around the wellbore, as shown in Table 3-2 and Figure 3-15.

3.4.3 Results of Calculation

3.4.3.1 Length changes of sanding disturbing zone, as shown in Figure 3-16

1. The drawdown is less than 2 MPa (from point a to point b), and the sanding disturbing zone does not occur or expand.
2. The drawdown is increased from 2 to 9 MPa (from point b to point c), and the sanding disturbing zone occurs and expands quickly.

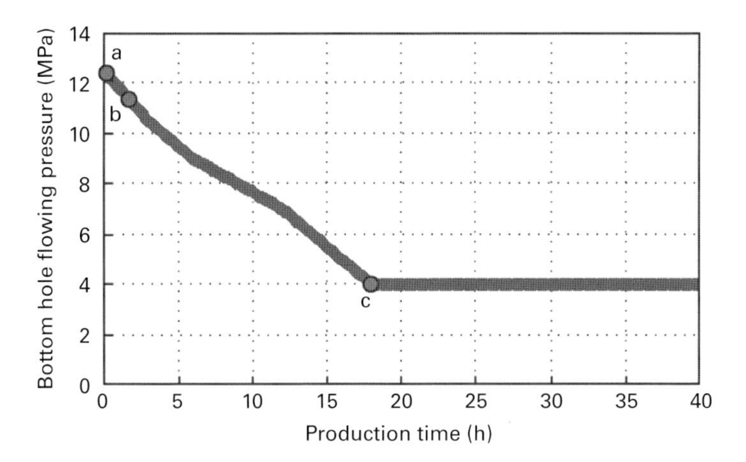

Figure 3-15 Bottom hole flowing pressure control of wells with limited sand production.

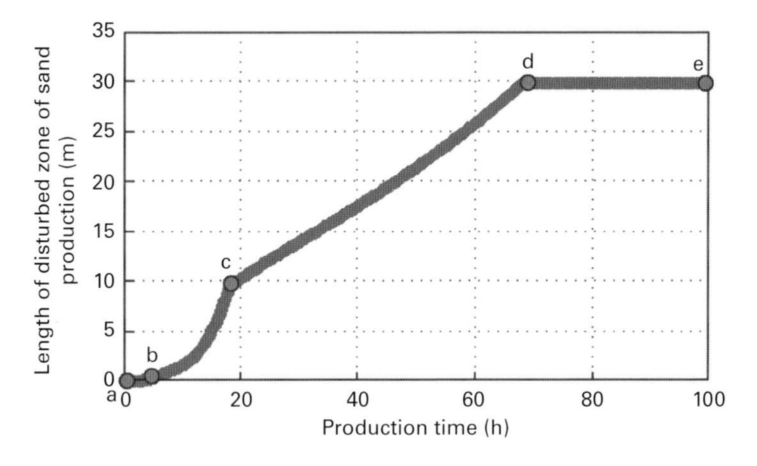

Figure 3-16 Length changes of sanding disturbed zone.

3. The bottom hole pressure remains constant after point c. The sanding disturbing zone will extend within the reservoir due to the delay effect of pressure transient, but the speed of extension becomes obviously slow and the pressure gradient lessens gradually.
4. The pressure gradient with the reservoir at point d is not enough for the sanding disturbed zone to grow. As such, the extension of this zone within reservoirs will stop.

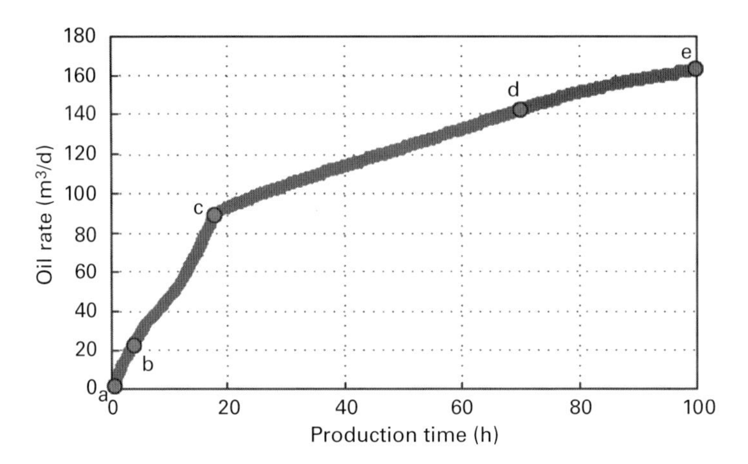

Figure 3-17 Changes of oil production rate.

3.4.3.2 Changes of oil rate, as shown in Figure 3-17

The feature of oil production rate can be categorized into four stages on the basis of increasing ratio:

1. Crude oil begins to flow (from point a to point b).
2. Oil rate increases rapidly with the development of sanding disturbed zone (from point b to point c).
3. When the drawdown is constant from (point c to point d), the extension of sanding disturbed zone decreases and the increase of oil production rate slows down accordingly.
4. When the extension of sanding disturbed zone stops (from point d to point e), sand production due to denudation dominates. Oil production rate will increase continuously due to the improvement of formation properties, but the growing ratio tends to be stable.

3.4.3.3 Changes of sanding rate, as shown in Figure 3-18

Sanding velocity can also be categorized into four stages:

1. The volume of sand production is zero before the formation of sanding disturbed zone (from point a to point b).
2. The volume of sand production increases rapidly with the extension of sanding disturbing zone (from point b to point c).
3. The increase of sand production volume slows down due to the slowing growth of sanding disturbing zone when drawdown is stopped (from point c to point d).

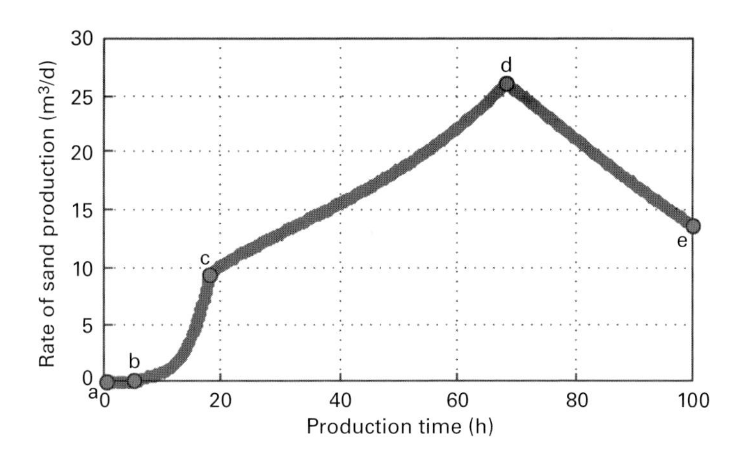

Figure 3-18 Changes of sand production rate.

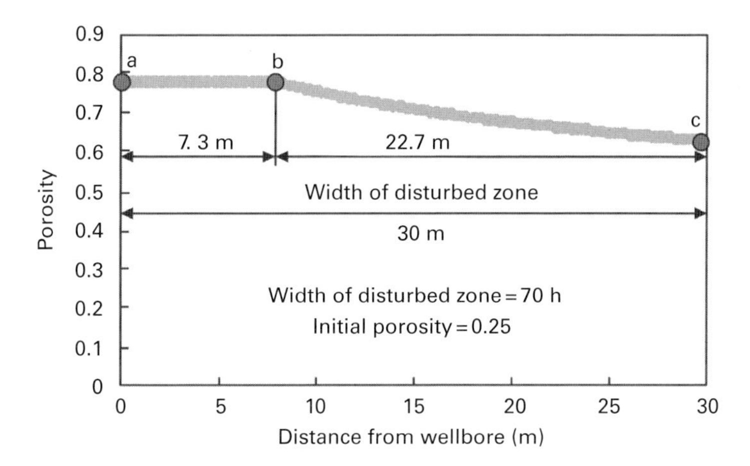

Figure 3-19 Porosity distribution within sanding disturbed zone.

4. Sand production rate decreases gradually when the extension of sanding disturbed zone stops and sand production due to denudation dominates.

3.4.3.4 Porosity changes of sanding disturbed zone, as shown in Figure 3-19

According to the simulation results, the porosity within the sanding disturbed zone is not always the same:

1. The erosion function ends from point a to point b, and the porosity reaches the maximum value within the sanding disturbed zone.

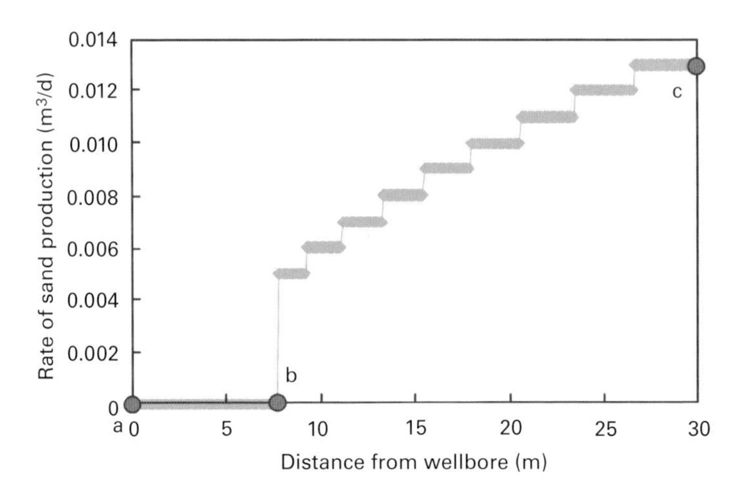

Figure 3-20 Distribution of sand production rate within sanding disturbed zone.

2. In the range between point b to point c, the area around point b is the disturbed zone resulting from both erosion and denudation. Erosion becomes weak gradually with the extension of sanding disturbed zone to inner reservoir, and only denudation exists in the front section of the disturbed zone.

3.4.3.5 Sand production rate along the sanding channels, as shown in Figure 3-20

The sand production rate along the sanding channels well indicates the sanding mechanism in different areas within sanding disturbed zone:

1. In the front end (0–7.5 m), the sand production rate is zero because the sanding disturbed zone is finished already.
2. In the middle section (7.5–30 m), both denudation and erosion take effect. The erosion weakens, denudation strengthens gradually and sanding rate increases. In the tip of channels, only denudation exists and the sand production rate is high.

4 Well Completion Approaches of Sanding Management

Well completion means building a reasonable channel between oil and gas formations and wellbores on the basis of geological characteristics of oil and gas pays and the technical requirements of development. To effectively develop oilfields and gasfields and extend life of oil and gas wells and maximize economical profits of oilfields and gasfields, the method of well completion should be selected based on the types of oil and gas reservoirs, the features of pays and applicable technical requirements. Only a reasonable type of completion will provide the best channel between oil and gas formations and wellbores, maximize the flowing area between formation and wellbore, and minimize the damage on pays and the flowing resistance to wellbore. As for sand production management, well completion includes effective control of sand production to protect the wellbore wall from collapsing and keep the production rate stable in the long run.

4.1 Methods of Excluding Sand Production

Sand production management is to exclude sand production selectively. Being somewhat different from conventional sand exclusion, sand control is still needed when applying sand production management. In this case, reasonable methods to control sand production should be selected based on the available sand production excluding technologies for sand production management.

Sand Production Management for Unconsolidated Sandstone Reservoirs, First Edition.
Shouwei Zhou and Fujie Sun.

4.1.1 Conventional Methods of Excluding Sand Production

In general, current techniques of controlling sand production can be categorized into two types: chemical sand production exclusion and mechanical sand production exclusion. Chemical sand production exclusion is used in a narrow scope of applications (mainly fine and silt sandstone formations). With the technical development of mechanical sand production exclusion, chemical sand production exclusion is less and less important and no longer considered as a commonly used approach of sand production. Currently, the most popular equipment/tools of excluding sand production around the world are as follows:

1. Slotted liners
2. Wire-wrapped screens
3. Prepacked screens
4. Premium screens
5. Gravel packs

4.1.2 Methods of Excluding Sand Production of Sand Production Management

After analyzing and comparing the above-mentioned well completion equipment/tools of sand production exclusion, it is proved that only wire-wrapped screens and slotted liners can control particle sizes of sand production precisely for the purpose of limited sand production. The mechanisms of controlling sand production using both tools are to prevent sand through rectangular seams. The difference between them is that the flowing area is larger for wire-wrapped screen and smaller for slotted liners. In practice, one of them can be selected according to the production rate of oil wells.

Previous studies find that only wire-wrapped screens and slotted liners can control particle sizes of sand production precisely to meet the requirements of allowing limited sand production. But the application cases show that sand particles will erode wire-wrapped screens and slotted liners, make seams of screens and liners widen, and thus affect the result of sand production exclusion. Even erosion damage of screens and liners may occur, resulting in high sand production rate.

Tools of excluding sand production applied in offshore oilfields include two categories, namely, gravel packs and premium screens. In order to apply sand production management successfully, three preliminary plans are proposed for further study on the basis of analyzing previous applications. These plans include choosing reasonable premium screens, enlarging allowed particle sizes through premium screen appropriately, and enlarging the size of gravels.

There are many types of premium screens. On-site trials in some oilfields of Bohai Bay used metal mesh and metal cotton premium screens in recent years, which have different structural features and conditions for application. The multilayer composite structure of conventional premium screens makes space blocking occur easily and the removal of blocking difficult. Planar structures are applied for metal cotton premium screens. When production drawdown is large enough, some fine particles can flow into wellbore to avoid blocking screens. So, the tidiness of screens can be realized by controlling production drawdown.

Metal mesh filter media has the following disadvantages.

1. With narrow applicable scope due to its mesh planar structure, being sensitive to distribution scope of formation sands (e.g. either excluding sand production completely or making massive sand production).
2. Easy blocking of screens, where fine particles can pass easily but larger particles or right sized particles will block mesh spaces after deforming, causing flowing areas smaller and smaller over time.
3. Meshes of screens will be enlarged due to erosion, affecting accuracy of controlling sand production.
4. With larger planar flowing areas, and without three-dimensional flowing areas.

The three-dimensional structure of metal cotton filter media has the following features.

1. With wide scope of applications, less sensitivity to particle size distribution of formation, and greater range of excluding sand particles.
2. Self-cleaning. Fine particles pass easily. Even blocking occurs, the planar flow path will not be blocked.
3. Self-healing. Parts of large particles will be squeezed out under the condition of high production drawdown. Screens can easily restore their original shape due to elasticity of metal wires.
4. Probably with smaller plane flowing area, but with larger three dimensional flowing area.

On the basis of analyzing applications of oilfields, it is found that screens of metal cotton filtering media are better than premium screens of metal mesh filtering media in excluding sand production.

Laboratory experiments studied applications of increasing accuracy of excluding sand production by premium screens and by enlarging gravel sizes of gravel pack.

4.2 Design of Sand Exclusion Parameters

4.2.1 Designing Aperture Width of Sand Production Exclusion Screens

The mechanism of sand production exclusion of wire-wrapped screens and slotted liner is to exclude larger particles from entering the wellbore by controlling seam width. Therefore, sand bridge will be formed by lager particles outside the screen, and sand production exclusion can be achieved, as shown in Figure 4-1. So, seam width of sand production excluding screens is the major parameter of this design.

The design of aperture width (e) at present is based on complete exclusion of sand production, and the methods used are as follows.

1. $e = d_{10}$, currently in use in US Gulf coastal areas (d_x represents the particle size of sieve analysis curve corresponding to a given $x\%$ cumulative weight).
2. $e = 2d_{10}$ (in use in California area) and $e \leq 2d_{10}$ (in use in China).
3. Con-slot considers the impact of sand sorting coefficient when designing aperture, and the methods of selection are:
   If $C < 2$, $e = d_{50}$,
   If $C = 2$, $e = d_{40}$,
   If $C > 2$, $e = d_{30}$.
   Where $C = d_{40}/d_{90}$, sorting coefficient.

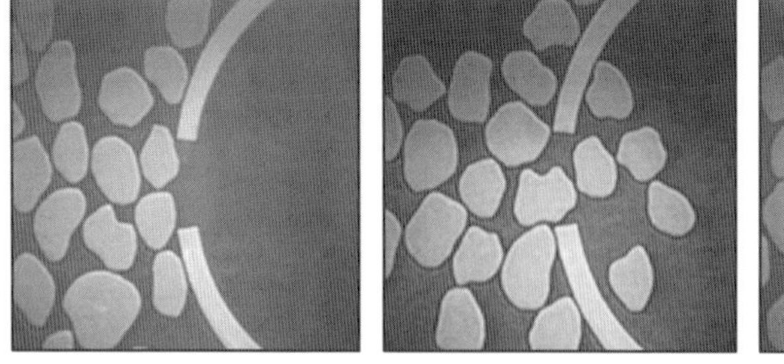

Figure 4-1 Traditional theory of sand bridging.

4. Markestad considered various factors and built a mathematical model, obtaining four parameters through calculation: the maximum aperture width ***d−−***, where blockage occurs frequently; the minimum aperture width ***d−***, where blockage will not occur; the maximum aperture width *d+*, where sand production will not occur; and the minimum aperture width *d++*, where continuous sand production will occur. Considering sand production control, the chosen aperture width is between *d−* and *d+*.

In order to compare the calculation results of these designing methods, the real data on particle sizes of formation sands from one offshore heavy oil reservoir in China are used for analysis and calculation (Figure 4-2).

The calculation result is 490 μm by using the method of d_{10}, 980 μm by using the method of two times d_{10}, 180 μm by using the method of Con-slot, and 200 μm by the method of Markestad. By comparing the aforementioned methods, it is found that the calculation results from various designing methods are different and the selection of a specific calculating method is very difficult (Figure 4-3).

The basic principal of controlling sand production by using rectangular apertures is to form sand arches outside rectangular aperture to prevent sand production. Coberly found through experiments that spherical particles cannot pass through rectangular apertures that are 2 times the particle size. Penberthy pointed out that spherical particles that are less than 2.5 times the

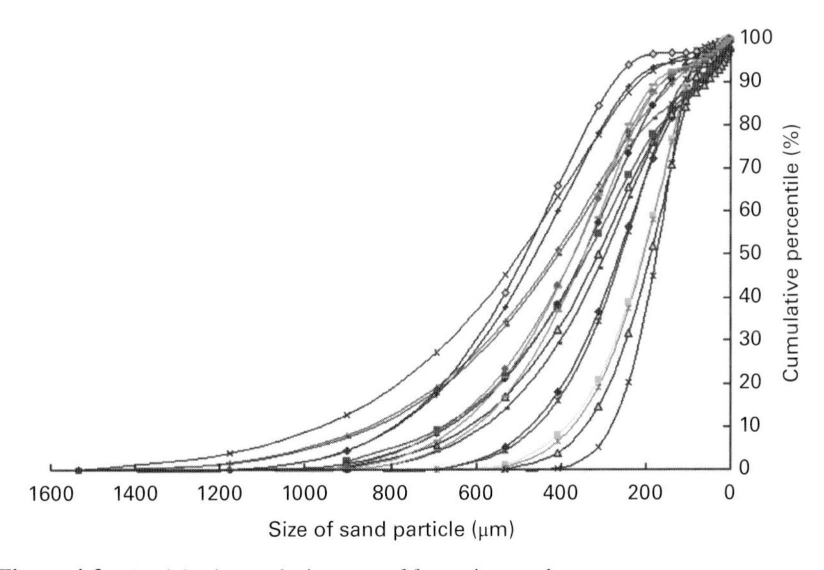

Figure 4-2 Particle size analysis curve of formation sands.

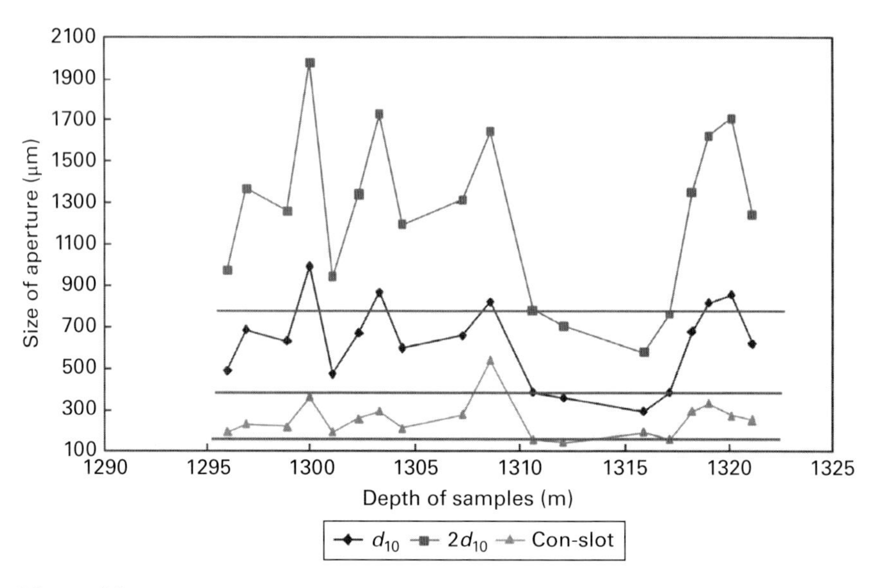

Figure 4-3 Designing results of apertures for normal sand control models.

rectangular aperture width will form an unstable bridge that will collapse when applying drawdown, causing massive sand production. So the size of apertures should be 2 to 2.5 times of bridging sand particles. If using d_{10} as bridging particle size, numerous formation sands whose size is less than d_{10} will pass through screens before formation sands of d_{10} particle size form a sand bridge, leading to poor sand controlling results. In order for better effects of preventing sand production, it is necessary to choose smaller particle sizes, like selecting particle size of d_{90} to control sand production from conservative standpoint to achieve the objective of sand exclusion. The calculation result by using 2 to 2.5 times of d_{90} is 280 μm. Compared with calculation results by other designing methods, this result is very close to that by Markestad method, indicating that Markestad method is a conservative approach to conduct estimation (Figure 4-4).

Therefore, from the view of safety and sand exclusion, we can use 2 to 2.5 times d_{90} and the Markestad approach to calculate aperture width, and enlarge aperture width appropriately based on the effect of applications in fields (Figure 4-5).

In order to allowing limited sand production, it is necessary to enlarge sand exclusion criteria appropriately on the basis of complete sand exclusion. Based on the definition of four aperture width of the Markestad approach, it is not difficult to find that the selected aperture sizes for achieving limited sand production should be between *d+* (the maximum aperture width that

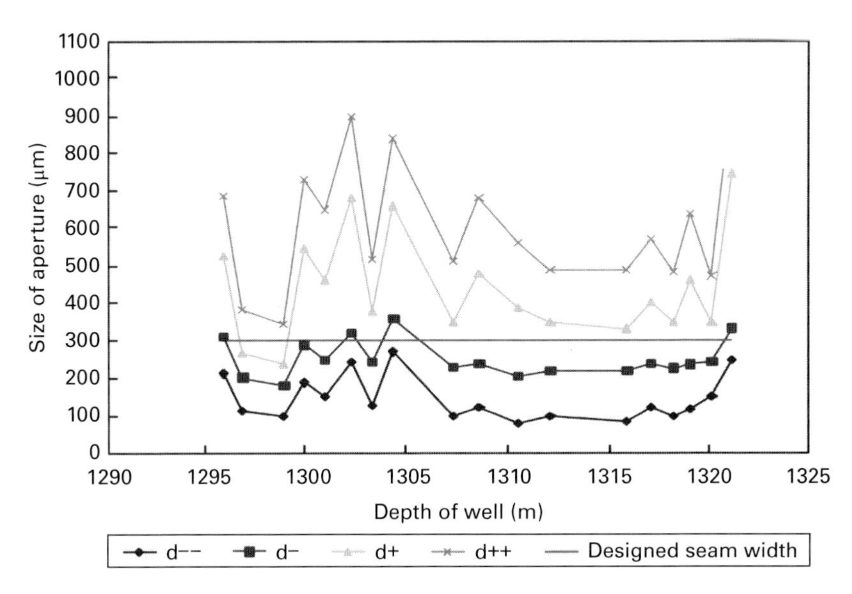

Figure 4-4 Designing result by the Markestad approach.

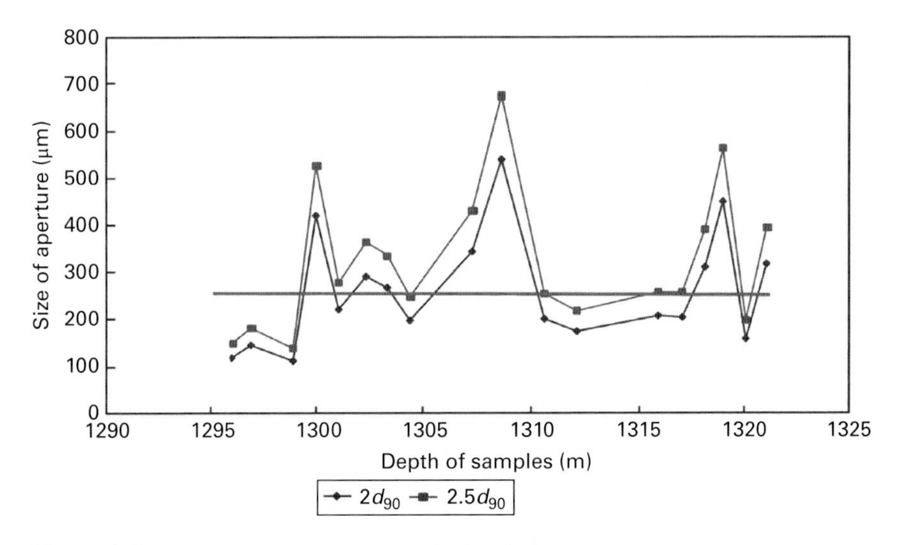

Figure 4-5 Result no.1 achieved from bridging design.

does not allow sand production) and *d*++ (the minimum aperture width where continuous sand production occurs). According to the calculating method of excluding sand production by rectangular aperture, particle sizes of sand exclusion should be enlarged (e.g. d_{85}, d_{90}, and so on). The selection of particle sizes of sand exclusion depends on sand production rate and surface

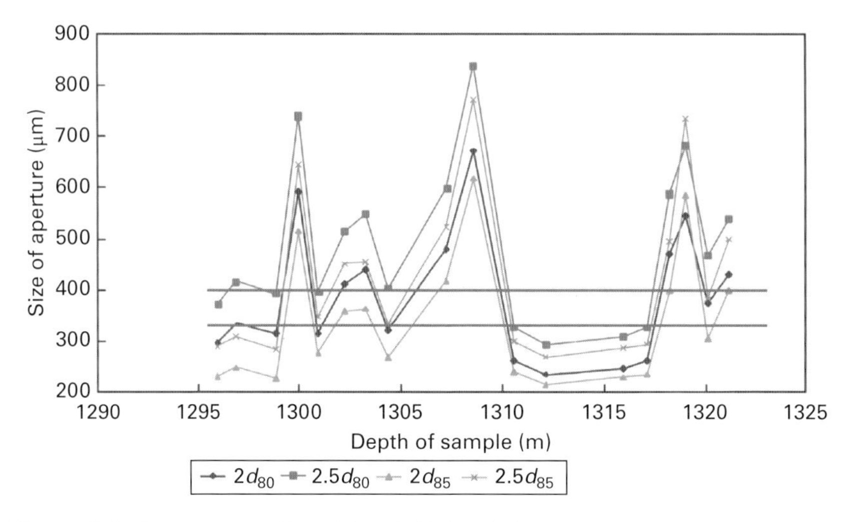

Figure 4-6 Result no. 2 achieved from bridging design.

sand-processing capacity. On the basis of preliminary result of sand production rate prediction of the oilfield, if particle sizes of sand exclusion are fixed at d_{85} and d_{80}, the sand concentration into wellbore will be 2.91% and 3.88%, respectively. The calculation supposes that sands less than particle size of sand exclusion can all be produced; in reality however, some sands less than particle size of sand exclusion will not be produced due to bridging effect. So, the actual sand concentration will be lower than the estimated value. Thus, the screen aperture can be enlarged appropriately as long as allowed by surface processing capacity (Figure 4-6).

The calculation results by these two methods are similar (e.g. 350 to 400 μm), and there is a certain increase when comparing with the calculation value of 280 to 300 μm based on the sand exclusion model. These results can be applied in the design of aperture width when allowing limited sand production.

4.2.2 Designing Methods of Sand Exclusion Precision for Premium Screens

In recent years, the technique of premium screens developed very fast and different manufacturers produced various types of products. Premium screens can be categorized into metal mesh premium screens and metal cotton premium screens based on sand exclusion mechanism. Generally, premium screens used for sand exclusion are adapted to formations with uniform sand distribution and coarse particle sizes.

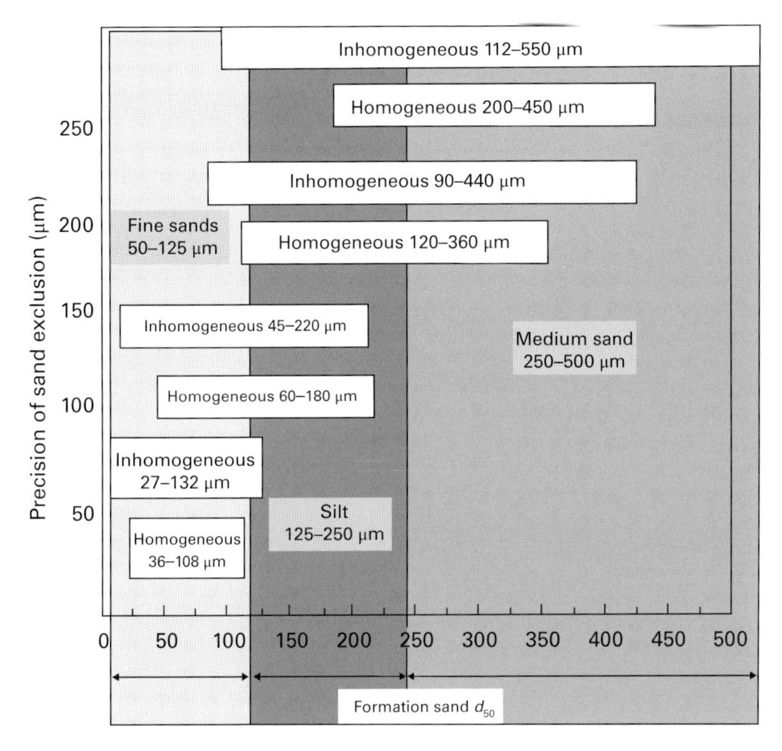

Figure 4-7 Relationship between mesh of standard metal mesh premium screens and formation sand size range.

At present, Saucier, Coberly, Schwartz, and colleagues recommended different aperture width design criteria. These standards, however, are limited to certain conditions and are hardly applicable premium screens developed in recent years. George Gillespie conducted lots of experiments by using metal mesh screen premium screens, including simulating sand samples of different particle sizes and heterogeneity, measuring sand production rate, sand particle sizes and pressure drop on two ends of screen when passing through standard mesh screens, through which he proposed the selection method of mesh of standard metal mesh premium screens (Figure 4-7). This approach takes the range of median formation sand size and the heterogeneity of pay into account, and the mesh diameter of screens is approximately 0.8 to 1.2 times of formation median d_{50}. The approach does not use actual screens to conduct experiments, so the achieved selection criteria is in a large range and cannot help to determine sand exclusion precision of screens accurately. Therefore, the method is limited in a certain degree when allowing limited sand production during development and it is necessary to conduct further research by laboratory experiments.

4.2.3 Approach of Gravel-packing

The reasonable match between packing gravels and formation sand is essential for sand exclusion and improving well productivity. In order to understand the relationship between packing gravels and formation sand, we suppose that gravels are uniform and spherical particles of the same size. There are two cases for overlapping spherical sections, of which one is of rectangular structure and the other is of triangular structure, as shown in Figure 4-8.

The black spheres above are packing gravels, and its diameter is D. The relationship between the diameter of gravels (D) and the inscribed circle diameter (d) for the two aforementioned cases through geometrical analysis and calculation are as follows:

For rectangular structure, $D = 2\left(\sqrt{2}+1\right)d \approx 5d$

For triangular structure, $D = \sqrt{3}\left(2+\sqrt{3}\right)d \approx 6d$

The studies show that, when the largest packing gravels are arranged with the first structure (i.e. rectangular structure), the formed pores are the largest in the gravel-packing zone. When the smallest packing gravels are arranged with the second structure (i.e. triangular structure), the formed pores are the smallest in the gravel-packing zone. As such, the recommended smallest diameter of gravels is 5 times the diameter of the largest allowable sand particles, and the suggested largest diameter of gravels is 6 times the diameter of the smallest allowable sand particles.

Saucier studied the relationship between permeability of the packing zone and the ratio (d_{50}) of median gravel medium sizes to medium formation sand sizes, as show in Figure 4-9. When the ratio d_{50} is 14, formation sand will flow through the packing zone freely. Due to high permeability of the packing zone, it is difficult for formation sand to form a "sand bridge" in the packing zone and massive sand production will occur. When the ratio d_{50} is 6 to 14, sand formation will flow into the packing zone, reducing the permeability. When the ratio d_{50} is 5 to 6, the passing of formation sand into the packing

Figure 4-8 Structures of packing gravels.

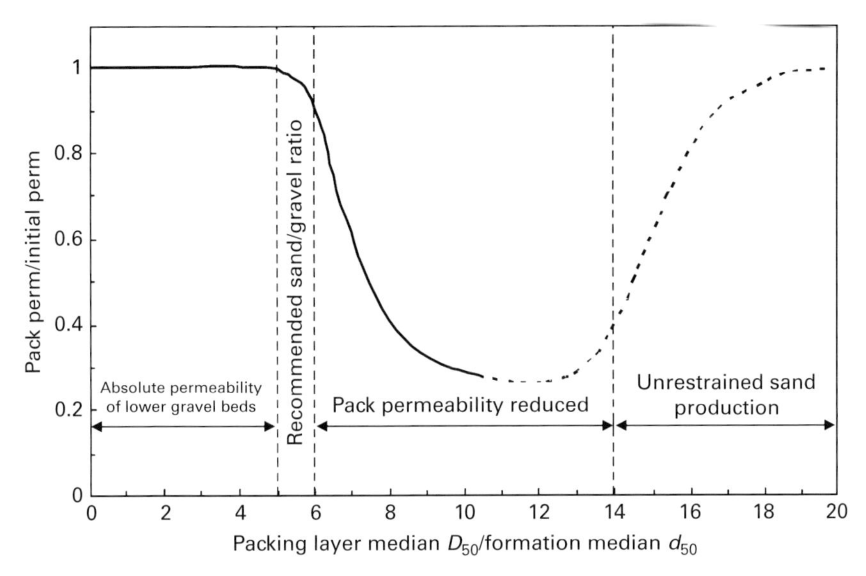

Figure 4-9 Empirical model of gravel size selection when excluding sand production by gravel-packing in fields.

zone will be prevented. Therefore, Saucier recommended that the ratio of median gravel medium size to median formation sand size is 5 to 6.

The approach proposed by Saucier is based on the experiments of gravels in ideal conditions, without considering such factors as gravels being under pressure when packing gravels in fields, heterogeneity of industrial gravels, and so on. So the method needs to be improved further.

4.3 Physical Simulation of Evaluating Sand Control Results

The current approaches of evaluating performance of sand pack screens worldwide are limited to study the mechanism of controlling sand production. There are no detailed plans and devices to estimate the effect of excluding sand production in field applications on the basis of engineering methods. According to the features of sand exclusion in unconsolidated sandstone heavy oil reservoirs, the full-scale wellbore testing equipment are studied and developed to simulate sand production/sand exclusion, which compensate for the shortcomings of small-scale appraising experimental methods and improve the appraisal study of sand exclusion screens from mechanism research to engineering evaluation stage.

4.3.1 Experimental Devices

These experimental equipment can be used to conduct simulation experiments of sand exclusion results under different methods of sand control, to evaluate advantages and disadvantages of sand exclusion approaches, and to choose the best measures of excluding sand, types of sand exclusion screens, parameters for sand exclusion and gravel size of gravel-packing.

The device consists of three components: simulation system, circulation system, and measurement system, as shown in Figure 4-10. Experimental fluid flows into the high pressure autoclave body from four directions under set pressure by a using high pressure pump, forming uniform radial flow through distributary mesh on the inner wall of autoclave body, passing through simulating formation sand and then carrying sand and flowing into annulus of simulation wellbore. A part of fine grain sand will be carried out through sand exclusion screens by fluid, but most of sands will accumulate outside sand exclusion screens and form sand bridges.

A unique feature of the device is that it can conduct evaluation experiments of sand production laws of radial flow with confining pressure, changing the past approach of study sand production law under the condition of axial flowing in small-scale cores, and making the experimental features of sand production much similar with the actual formation conditions.

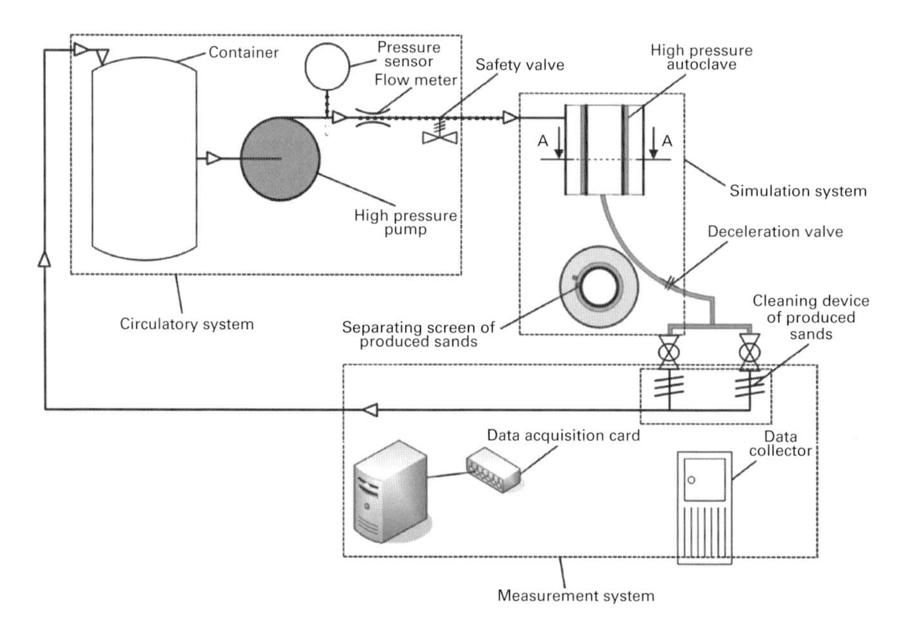

Figure 4-10 Experimental devices of sand exclusion.

4.3.1.1 Simulation system

As for the sand packing container among the sand production simulation devices for full-size wellbore, the height is 340 mm, inner diameter is 430 mm, and outer diameter is 460 mm, as shown in Figure 4-11. It is made of stainless steel, resistant to high pressure and corrosion, and the maximum working pressure is up to 40 Mpa.

When conducting experiments, a bracket supporting wellbore wall is placed between sand exclusion screens and simulation formation sand, consisting of a stainless steel tube with punctured holes. Its outer diameter is 215.9 mm (adjustable), used to support weakly consolidated sand. Sand exclusion screens, the bracket supporting wellbore wall and simulation formation sand are placed from inner to outer in sequence (artificial simulation sand formation is compacted). During the experiment, rubber pads are used to ensure the container sealed.

In order to better simulate downhole radial flow, four oil injection inlets are evenly distributed in sidewall surface of sand packing container as radial oil injection ports (Figure 4-12).

4.3.1.2 Circulation system

Circulation system of the device includes fluid pump, pressure control system, high pressure line, fluid diverter valve, and so on. The fluid pump

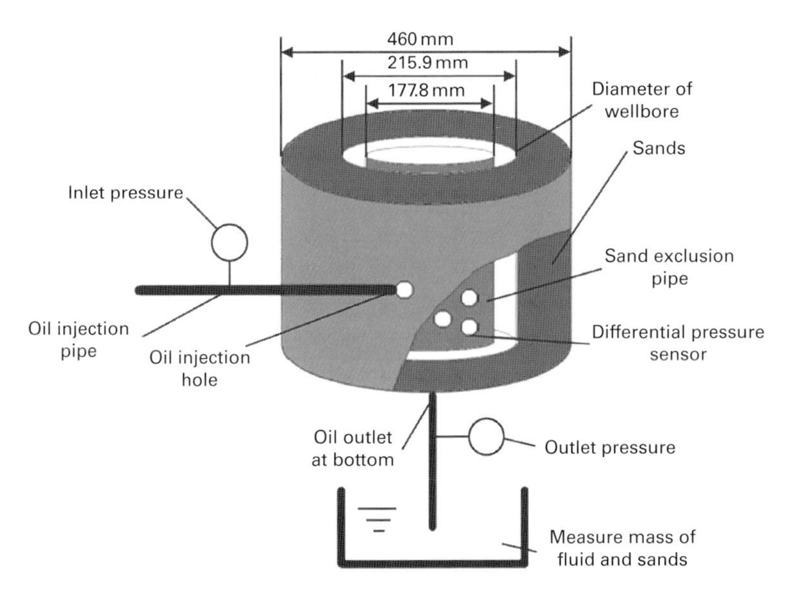

Figure 4-11 Conceptual experimental devices.

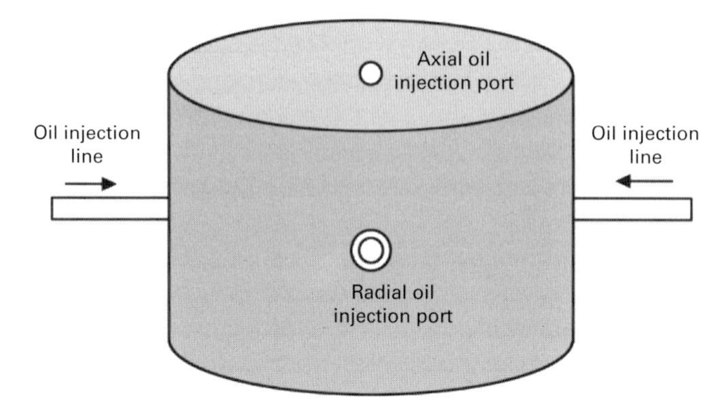

Figure 4-12 Oil injection port.

provides liquid for sand packing container with constant pressure to simulate the drawdown of formation. The fluid used in the experiment is oil for fracturing, with viscosity of 50 to 300 mPa·s. The circulating line is a stainless steel pipe with 3-mm outer diameter and 1.5-mm inner diameter. The liquid and sand separator are installed under the outlet ball valve. When the oil flows out from the container outlet, the separated oil will be reused.

4.3.1.3 Measurement system

According to the requirements of experiments, the pressure measuring equipment are mounted in injection port, outside of sand exclusion screens, and in the middle of annulus respectively. The pressure measuring points shown in Figure 4-10 are used to monitor the change of pressure. In the outlets, measuring and weighing are performed continuously for fluid with sand content. At the same time, a sieve with 400 mesh is used to separate sand and liquid according to the distribution range of formation grain size (the opening size of 400 mesh sieve is 0.004 mm, which can capture all produced sand). The gathered sand is then dried. The method of separating sands from produced liquid continuously in a certain period of time can measure the amount of continuous sand production within this period.

4.3.2 Experimental Evaluation Index

In order to better reflect the realities in fields, the parameter of "productivity index per meter" is used as the evaluation reference of this experiment.

The productivity index per meter refers to the production rate per unit pressure and per unit thickness, so the equation is as follows:

$$PI_m = \frac{Q}{\Delta p \cdot L} \tag{4-1}$$

Where PI_m is productivity index per meter, $m^3/(MPa \cdot m \cdot d)$; Q is production rate through sand exclusion screens or gravel-packing formation, m^3/d; L is effective flowing length of sand exclusion screens or gravel-packing formation, m; and Δp is production drawdown, MPa.

The pressure difference between the liquid pressure and outlet pressure does not represent practical production drawdown in this experiment. In this case, we need to acquire the pressure of actual supplying radius by using pressure distribution equation of the radial flowing through testing pressure gradient of simulation formation interval, plus the additional pressure drop nearby wellbore screen or gravel-packing zone, by which well drawdown is calculated and productivity index per meter is achieved. The experimental result is supposed to better conform to practical conditions in the field.

$$\Delta p = \Delta p_1 + \Delta p_2 \tag{4-2}$$

Where Δp is production drawdown, MPa; Δp_1 is pressure drop of testing screen or gravel pack layer, MPa; and Δp_2 is real pressure drop in formation, MPa.

4.3.3 Evaluation Experiments of Sand Exclusion Precision of Premium Screens

4.3.3.1 Experimental conditions

According to distribution features of formation particle sizes of an offshore oilfield in China, Particle size eigenvalues of quartz sand for experimental simulation are determined as shown in Table 4-1. Pressure difference of provided liquid is 3 MPa; fluid viscosity is 200 mPa·s; and parameters of sand exclusion of standard metal mesh screens are 50, 100, 150, 230, and 300 μm.

Table 4-1 Particle Size Eigenvalues of Experimental Sands

Eigenvalues of Sand Particles	d_{10}	d_{40}	d_{50}	d_{90}	*UC*
Quartz sand for experiments (μm)	363.6	221.2	170.1	50.3	4.4

4.3.3.2 Experimental processes

First place metal mesh premium screens of 50-μm sand exclusion parameter in autoclave, fill the pressure container with simulation quartz sand, set working pressure of liquid providing pump at 3 MPa, then start to increase pressure and circulate the liquid, and measure pressure, flow rate and sand production rate of all test points. Stop the experiment when flow rate and pressure become stable. Then dismantle the apparatus, and change the parameters of sand exclusion screens to 100, 150, 230, and 300 μm, respectively, and repeat the experiment.

4.3.3.3 Experimental analysis

Through measuring flow rate, pressure drop and sand production rate of five groups of experiments, productivity index per meter, and sand concentration in oil are calculated as shown in Table 4-2.

A diagram is drawn based on the data of productivity index per meter and sand concentration, providing reference for optimizing the design of sand exclusion precision of premium screens, as shown in Figure 4-13.

Conducting regression of measuring data can lead to the regressive relationship between the three parameters:

$$C_{\text{sand}} = 0.5048(\omega / d_{50}) - 0.1419 \tag{4-3}$$

$$PI_m = 1.2152 \ln(\omega / d_{50}) + 1.6501 \tag{4-4}$$

4.3.3.4 Experimental results

Conventional sand exclusion design for offshore fields requires that sand concentration should be controlled under 0.3%. Substitute this parameter in

Table 4-2 Experimental Data of Sand Production Simulation Using Metal Mesh Premium Screen (Experimental Sand d_{50} = 170.1 μm)

Excluding Parameter ω (μm)	*ω/d_{50}*	*Flow Rate (×10^{-3} m^3/min)*	*Pre. Drop of Screen Δp1 (MPa)*	*Real Pre. Drop Δp (MPa)*	*PIm [m^3/(MPa·m·d)]*	*Sand Concentration (%)*
50	0.29	0.19	2.80	6.65	0.27	0.060
100	0.65	0.59	2.65	6.50	0.87	0.086
150	0.88	0.95	2.60	6.45	1.41	0.377
230	1.32	1.353	2.50	6.35	2.04	0.535
300	1.76	1.59	2.45	6.30	2.42	0.740

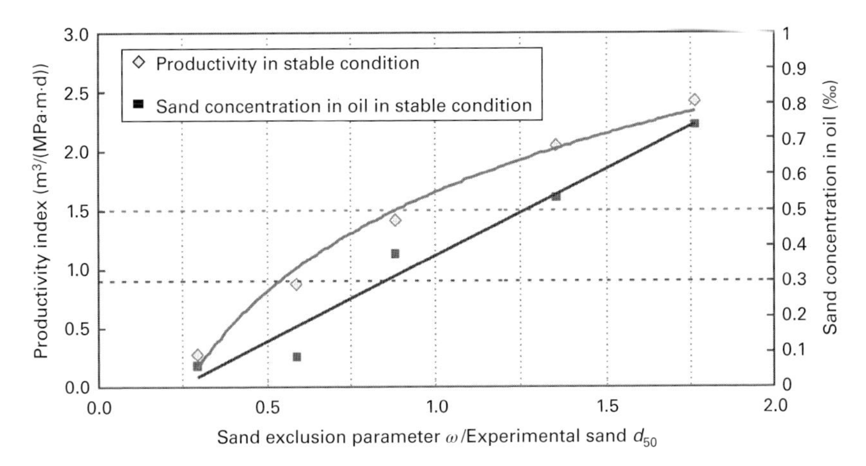

Figure 4-13 Experimental result diagram of evaluating woven mesh premium screens.

Equation (4-3), the relationship between sand exclusion parameter of screens and median sand particle size of formation can be achieved:

$$\omega \leqslant 0.88\ d_{50} \tag{4-5}$$

Allowing limited sand production when producing can improve well productivity, so sand exclusion parameters of screens can be enlarged appropriately on the basis of conventional sand exclusion design. According to the established relationship, when sand concentration is controlled within 0.5%, sand exclusion parameter of screens can be enlarged to:

$$\omega \leqslant 1.27\ d_{50} \tag{4-6}$$

Based on the above-mentioned conclusion, when substituting Equations (4-5) and (4-6) into Equation (4-4), the change ratio of productivity index of allowing limited sand production over productivity index of conventional sand exclusion design can be calculated:

$$\Delta PI = \left(PI_1 - PI_2\right) / PI_2 \times 100\% = 30\%$$

4.3.4 *Evaluation Experiments of Packing Gravel Sizes*

Currently, the commonly used practical method for designing gravel-packing parameters is the Saucier approach, where the gravel size is 5 to 6 times of the median sand particle size d_{50}. The approach proposed by Saucier is based on the experiments of ideal gravels, without considering such factors as gravels

Table 4-3 Particle Size Eigenvalues of Experimental Sands

Eigenvalues of Sand Particles	d_{10}	d_{40}	d_{50}	d_{90}	*UC*
Quartz sand for experiments (μm)	463. 6	281.2	220	70.5	4.0

being under pressure when packing gravels in fields, heterogeneity of industrial gravels, and so on. Considering the current model of allowing limited sand production, the issues such as whether results from this method are conservative, whether gravel sizes can be enlarged, and whether the sand production rate is within the allowable range need to be addressed by conducting estimation experiments of gravel-packing sand exclusion.

4.3.4.1 Experimental conditions

According to distribution features of formation particle sizes of an offshore oilfield in China, particle size eigenvalues of quartz sand for experimental simulation are determined as shown in Table 4-3. The liquid pressure difference is 3 MPa; fluid viscosity is 200 mPa·s. The sizes of packing gravels are 20 to 40 mesh, 10 to 30 mesh, 10 to 16 mesh, and 8 to 12 mesh. In the case of installing 5.5-in. sand exclusion pipe in 8.5 inches wellbore, the calculated gravel-packing thickness is 30 mm.

4.3.4.2 Experimental processes

Place wire-wrapped screens in pressure autoclave, fill 20 to 40 mesh industrial gravels outside screens, and fill the pressure container with simulation quartz sand. Set the working pressure of liquid providing pump at 3 MPa, then start to increase pressure and circulate the liquid, and measure the pressure, flow rate, and sand production rate of all test points. Stop the experiment when flow rate and pressure becomes stable. Then dismantle the apparatus, and change the gravel sizes and repeat the experiment.

4.3.4.3 Experimental analysis

Through measuring flow rate, pressure drop, and sand production rate of four groups of experiments, productivity index per meter and sand concentration in oil can be calculated as shown in Table 4-4 and Figure 4-14.

4.3.4.4 Experimental results

1. When $D_{50}/d_{50} < 6$, sand concentration is basically within 0.2%. Good result of sand exclusion can be achieved, and productivity index per meter increases with the enlargement of gravel sizes.

Table 4-4 Experimental Data of Sand Production Simulation Using Metal Mesh Premium Screen (Experimental Sand d_{50} = 220 μm)

Mesh of Gravel	D_{50}/d_{50}	*Flow Rate ($\times 10^{-3}$ m^3/min)*	*Pre. Drop through Gravel $\Delta p1$ (MPa)*	*Real Pre. Drop Δp (MPa)*	*PIm [m^3/(MPa·m·d)]*	*Sand Concentration (%)*
20 to 40	2.9	2.00	1.30	5.15	3.7254	0.15
10 to 30	5.9	2.13	0.68	4.53	4.5102	0.18
10 to 16	7.3	2.50	0.66	4.51	5.3171	0.60
8 to 12	9.2	1.70	1.40	5.25	3.1063	1.00

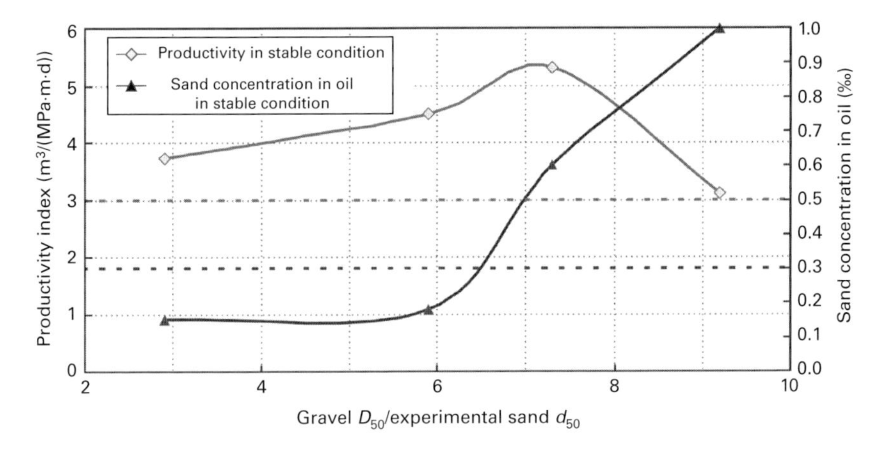

Figure 4-14 Experimental result diagram of evaluating gravel-packing.

2. When D_{50}/d_{50} = 6.5, sand concentration reaches 0.2%. When D_{50}/d_{50} = 7, sand concentration reaches 0.5‰. The productivity index per meter increases with the enlargement of gravel sizes.
3. When D_{50}/d_{50} > 7, the productivity index per meter begins to decrease. When the ratio reaches 9.2, the productivity index per meter decreases by 42% compared with peak value, and sand concentration increases by nearly 70% as compared with that of peak PI per meter, indicating that serious sand encroachment occurs in the gavel packing zone and some of fine sands are produced by flowing through sand packing zone and sand exclusion screens.
4. In the case of allowing limited sand production, if allowable sand concentration is controlled under 0.5%, the gravel sizes can be enlarged to about 6 to 7 times D_{50}/d_{50}, and the productivity index improves by 18% to 20% compared with classic theory of sand exclusion proposed by Saucier.

5

Sand Carrying in Wellbore and Surface Treatment of Produced Sands

In order to develop unconsolidated sandstone reservoirs effectively, fine particles eroded nearby wellbore should be evacuated to enlarge flowing channels in reservoirs, improve permeability and reduce flowing resistance. To realize this purpose, the produced fine particles from formation should be smoothly carried to surface as much as possible with the flowing of well effluent. If produced fine particles deposit in the bottom of wells, tubing, artificial lift facilities, and downhole tools, they will block the flowing channels, erode mechanical equipment and downhole tools, and result in the loss of production time of wells. For offshore oilfields, the fine particles carried to surface need to be separated by using oil–water separator, and these particles contained with oil after treatment should meet the environmental discharge requirements. During the development of oilfields, producing conditions and sand production status of wells are changing over time, so sand production surveillance is needed to estimate the situation of sand production, adjust related operating parameters in time, make production rate, capability of carrying sand along wellbore, and surface treatment capacity match well, and thus reach the purpose of optimizing production.

5.1 Critical Flowing Velocity of Carrying Sand in Wellbore when Producing with Sands

The goal to study the migration law of sand particles in wellbore is to optimize production rate to carry produced sand, reduce the sand settling and accumulation in wellbore, extend the cycle of flowing produced sand back, and

Sand Production Management for Unconsolidated Sandstone Reservoirs, First Edition.
Shouwei Zhou and Fujie Sun.

maximize economic profit. In order to accurately predict sand-carrying capacity in wellbore, lots of theoretical and experimental researches are conducted in the past, but there are still a few problems requiring further clarification. In theory, the past research mainly focuses on static aspects. The settling principles of solid particles in static fluid are actually already very perfect. What requires further study is the moving law under the condition of flowing. Currently, there are various calculating formulas with different expressions. However, most of them are not based on all related parameters and lead to large errors due to the difference from experimental conditions, and therefore they cannot be used universally. In the experimental aspect, as the experimental conditions like measuring geometrical condition of pipes, material, and manufacturing precision are not the same, they all will influence experimental results. It is very difficult to gather reliable and correct experimental data and to process the collected data because of many influential factors.

The prediction of sand-carrying issue in wellbore includes sand-carrying capacity and the calculation of pressure drop gradient. Actually, to identify sand-carrying capacity in wellbore is to calculate the critical sand-carrying flowing velocity, and the calculation of pressure drop gradient is to achieve pressure loss under the condition with different sand concentration, so the reasonable parameters of artificial lift equipment can be determined on the basis of these calculations. Due to the low sand concentration, to calculate pressure loss is based on weighted average of produced effluent density, so this aspect will not be discussed in this book.

Considering the above-mentioned difficulties, the following technical roadmap is used to strengthen theoretical supporting of research results on the condition of movement and reduce the level of experimental data processing. On the basis of current achievements, this chapter analyzes the forces acting on particles and moving conditions, builds theoretical model of particle movement in wellbore on the extreme conditions, improves theoretical model through experiments, and presents theoretical prediction of sand carrying in wellbore on the condition of producing.

5.1.1 Basic Theoretical Model

Figure 5-1 is a chart of forces analysis of solid particles in the slope of an inclined pipe, and the forces are gravity (F_g), frictional force (F_f), drag force (F_d), buoyancy (F_b), and so on. The movement in the pipe is impacted by particle properties, fluid properties, geometrical condition of pipe, external random factor, etc. During practical operations like daily production and well washing, individual solid particles do not exist independently due to

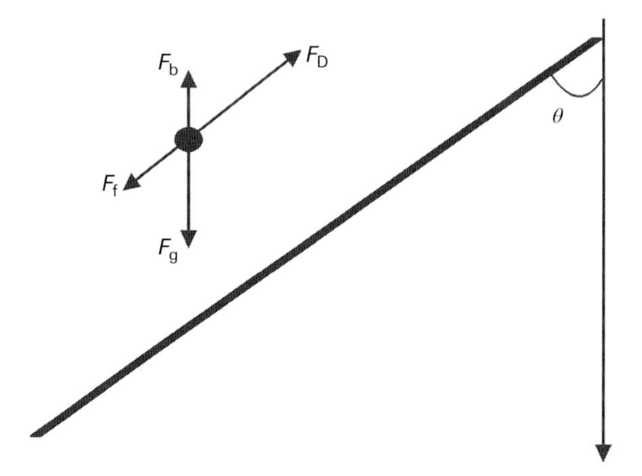

Figure 5-1 Forces on sand particles.

the limitation of tubing size. Especially for wells with serious sand production, the collisions among particles and between particle and tubing wall or casing wall will occur due to high sand concentration. When particles move in the pipes, if the sand-carrying velocity for certain particle size is constant, with the decrease of well inclination, the movement pattern changes gradually from jumping and moving bed mode to nonuniform, suspended liquid laminar flowing, and then transits to suspending liquid flowing. Particles with different sizes tend to transport in a minimal energy manner. When the diameter of particle is small, particles are tending to move in a laminar flowing as the critical transport condition; when the diameter of particle is large, particles are tending to move in a moving bed manner as the critical transport condition. In the case of practical sand carrying, the factors like particles rotation, external vibration, particle accumulation, etc. resulting from fluid velocity distribution will cause the actual drag force to decrease and the settling rate to accelerate.

To determine the critical sand-carrying capacity, we consider fluid properties, sand particle properties, pipe size, well inclination, interference of group of particles, effect of container wall, and so on, and use the method that is adaptable to all ranges of concentrations and Reynolds numbers.

When solid particles transport in moving bed mode, the formula of critical flowing velocity is:

$$u_c = \alpha \cdot u_d + 0.0285\sqrt{gD}\frac{\sin\theta}{1-\cos\theta} \tag{5-1}$$

When solid particles transport in laminar solid and liquid mode, the formula of critical flowing velocity is:

$$u_c = \alpha \cdot u_d \sqrt{(\cos\theta + \sin\theta \tan 65°)} \tag{5-2}$$

When solid particles transport in vertical pipes, the formula of critical flowing velocity is:

$$u_c = \sqrt{\frac{4(\rho_3 - \rho_1) g d_p \phi}{3\rho_1 C_D}} (1 - C_3)^n f_w \tag{5-3}$$

When sand particles transport in inclination wellbore in moving bed mode, the formula of critical flowing velocity is:

$$u_c = \sqrt{\frac{4(\rho_3 - \rho_1) g d_p \phi}{3\rho_1 C_D}} (1 - C_3)^n f_w + 0.0285\sqrt{gD} \frac{\sin\phi}{1 - \cos\phi} \tag{5-4}$$

When sand particles transport in laminar mode, the formula of critical flowing velocity is:

$$u_c = \sqrt{\frac{4(\rho_3 - \rho_1) g d_p \phi}{3\rho_1 C_D}} (1 - C_3)^n f_w \sqrt{(\cos\phi + \sin\theta \tan 65°)} \tag{5-5}$$

Where u_c is critical sand-carrying velocity, m/s; u_d is settling velocity of particles, normally regarded as interfering settling velocity, m/s; D is wellbore inner diameter, m; θ is tilting angle of transporting pipe or inclination, °; g is gravity, m/s^2; C_D is drag force coefficient of particle sedimentation; ρ_l is fluid density, kg/m^3; ρ_s is particle density, kg/m^3; α is correction factor; d_p is particle diameter, m; C_s is particle volumetric concentration, m^3/m^3; f_w is correction coefficient of container wall, as shown in Table 5-1; n is concentration interference coefficient, as shown in Table 5-1; and ϕ is shape factor of particles (i.e. ratio of surface area of sphere to that of the particle with the same volume).

As for transportation of sand particles in horizontal pipes, theoretical prediction result is that the moving particles at critical sand-carrying velocity in the horizontal wellbore will transport in moving bed mode, but practical calculation finds out that laminar flowing mode is more consistent with experimental results. The reason may be that the sands move in laminar flowing mode.

$$u_c = \sqrt{\frac{4(\rho_3 - \rho_1) g d_p \phi}{3\rho_1 C_D}} (1 - C_3)^n f_w \sqrt{\tan 65°} \tag{5-6}$$

Table 5-1 Calculation Method of the Values of n and f_w

Approach of Calculating n		*Approach of Calculating f_w*	
Formula of Calculating n	*Range of Reynolds*	*Formula of Calculating f_w*	*Range of Reynolds*
$n = 4.65 + 1.5\ (d_p/D)$	$Re_t < 0.2$	$1 - dp/D$ $1 + 0.475dp/D$	$Re_t < 2$
$n = (4.35 + 17.5^p)Re_1^{0.03}D$	$0.2 \leq Re_t < 1$	$f_p=1+235d/D$	$2 \leq Re_t < 500$
$n = (4.45 + 18d_p)Re^{-0.1}$	$1 \leq Re_t < 200$	$f = 1-(d/D)_{1.5}$	$500 \leq Re < 2 \times 10_5$
$n = 4.45Re^{-1.1}$	$200 \leq Re < 500$		
$n = 2.39$	$500 \leq Re_t < 2 \times 10_5$		

Note: Re_t is the settling Reynolds of particles.

5.1.2 Experimental Research of Sand Carrying in Wellbore

5.1.2.1 Introduction of experimental apparatus and materials

The experimental apparatus of sand carrying in wellbore is shown in Figure 5-2, including booster pumps for sand-carrying liquid, flow rate–controlling valves, flowmeters, formation sands, sand production rate–controlling valves, Plexiglas sand-carrying pipes, monitors of sand transportation, sand dropper for static settling, sand settling containers, sand recycling tanks, clean water tanks, flowing lines, and so on.

The liquid rate in offshore oilfields is usually high, so the size of tubing is commonly 2 in. and 3 in. and the inner diameter of tubing is 62 to 73 mm. The pipe with inner diameter of 65mm is used in the experiments. The pipe length is determined to be 4.2 m after considering the factors like inner diameter, the time of flow regime stabilization, fluid viscosity, pipe wall roughness, and so on.

In order to adjust experimental parameters, observe and detect the settling of sand particles, and meet the experimental requirements of horizontal pipe, vertical pipe and tilting pipe, the main flowing lines are using transparent Plexiglas to carry sand, and the tilting angle can be adjusted between 0° to 90° according to the experimental requirements. The throttle valve is used to control liquid rate. The experimental liquid is mainly clean water, the sand particles for experiments are formation sands processed by crushing and standard sieve analysis, and the average value of adjacent mesh aperture is used as parameter of experimental particle size, including seven particle sizes (e.g. 165, 192, 269.5, 392, 525, 750, and 1075 μm). According to the characteristics of particle size distribution of formation sand in Bohai Bay oilfields, we add sand particles into the pipe according to the modulus of about 1.0.

First add sand particles whose distribution of particle size is known into a device used to store experimental sands, and allow the liquid to flow through

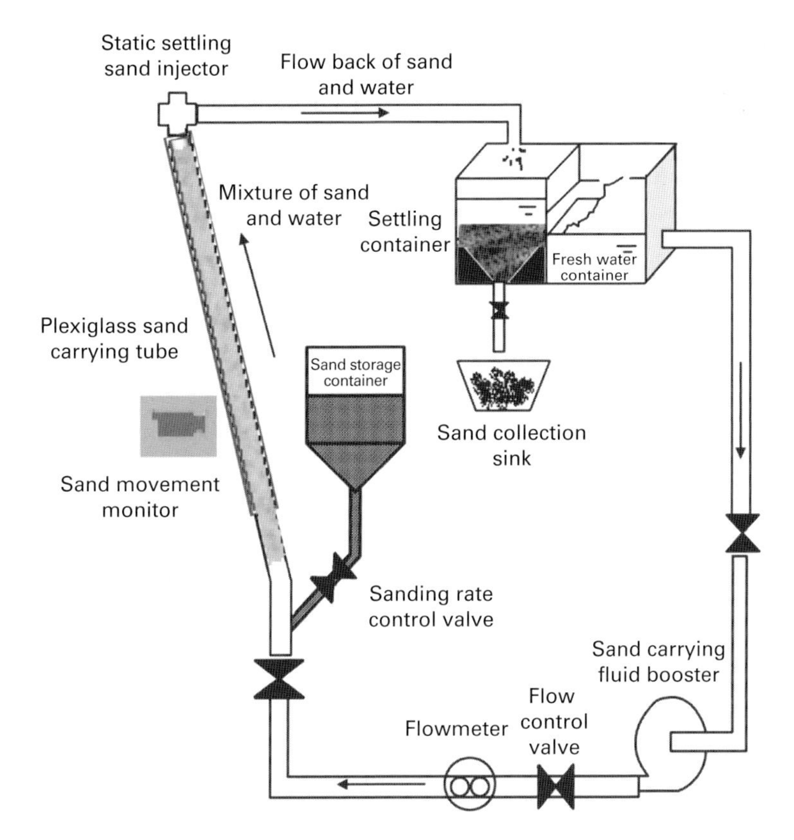

Figure 5-2 Experimental apparatus of sand carrying in wellbore.

pipe at a low flowing rate. Then increase flowing rate by using flow rate adjustment valves, carrying out sand particles according to the ascending order of particle sizes (from small to large). Use standard sieve to check the size of particles carried by flowing liquid at the outlet of sand-carrying pipe, and measure fluid rate at the same time. The critical sand-carrying velocity for different particle sizes can be achieved.

5.1.2.2 Laboratory experiments

1. Static settling experiments
   The purpose of static settling experiments is to obtain the shape factor of experimental sands. The basic experimental steps are as follows:

   1. Gently put single wetted sand particle into experimental pipe (Plexiglas) which is filled with experimental fluid.

Table 5-2 Shape Factors of Sand Particles

Diameter of Particle (μm)	*Real-Time Settling Velocity (m/s)*	*Free Settling Velocity (m/s)*	*Shape Coefficient of Particles*
525	0.0546	0.0743	0.541
383	0.0402	0.0518	0.603
270	0.0294	0.0348	0.718
192	0.0219	0.0250	0.771
165	0.0158	0.0198	0.638
Average value of shape coefficient of particles			0.654

2. Use a stopwatch to measure the time of sand particle passing through 0.5- to 1-m stable measuring section. Calculate an average value every 20 groups of data, and calculate the final velocity of settling of sand particles.
3. Calculate the shape factor on the basis of measured final velocity of settling and settling formulas.

Shape factors of experimental sands calculated by experimental results are shown in Table 5-2.

2. Sand-carrying experiments in wellbore
The purpose of this experiment is to measure sand-carrying capacity of fluids in different conditions like different tilting angle of pipe and different flowing velocity of fluid. The method of mathematical statistics is used to improve the theoretical model, and then critical migration principle of sand particles can be achieved on the conditions of different fluid flowing rate and wellbore geometry.
Experimental procedure is as follows:

1. Load sand particles with known distribution of particle size into the device which supports sand on the bottom of the Plexiglas pipe.
2. Adjust the angle of the Plexiglas pipe.
3. Start the fluid-pumping device, and adjust the controlling valve to increase flowing rate and flowing velocity in the experimental pipe when conducting the experiments.
4. Observe the movement of sand particles carefully. When sand particles of the smallest size (e.g. maximum standard sieve mesh: 100 mesh) appear, stop increasing fluid-flowing velocity and keep the flow rate stable, and measure flow rate of fluid in a certain time period. In order

to make the measurement of flow rate accurate, the measuring will commonly be repeated 3 to 4 times for a certain flowing rate.

5. Change sieve with larger holes and increase flowing rate. Repeat step (4) until the sand particles with the largest size are carried out.
6. Change sand samples, adjust the angle of sand-carrying pipe, and repeat steps (3) to (5).

3. Qualitative understanding of sand-carrying experiments
Hundreds of groups of sand-carrying experimental data were gathered at different tilting angle of pipe through experiments. With the help of imaging technology, flowing pattern pictures of sand particles at six various tilting angles (0°, 30°, 45°, 60°, 75°, and 90°) were obtained (shown in Figure 5-3 to Figure 5-8). Three flowing patterns of sand particles such as laminar flowing, jumping and moving bed flowing and uniform suspension flowing were observed (Table 5-3).

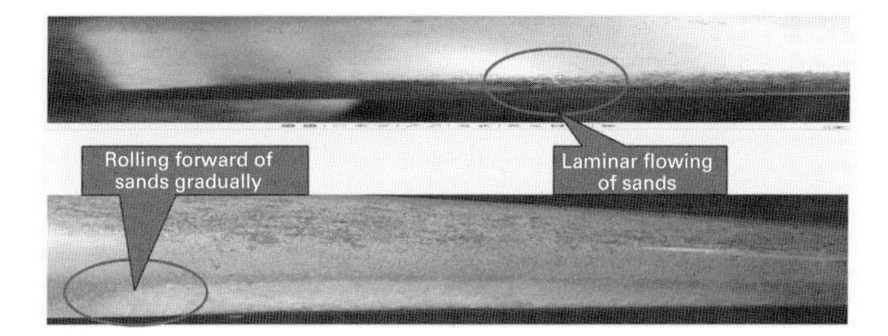

Figure 5-3 Picture of flowing of sand particles in horizontal pipe.

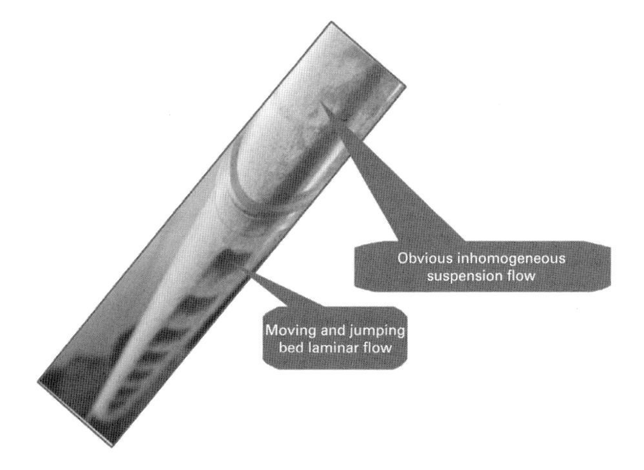

Figure 5-4 Picture of flowing of sand particles in tilting pipe at 30°.

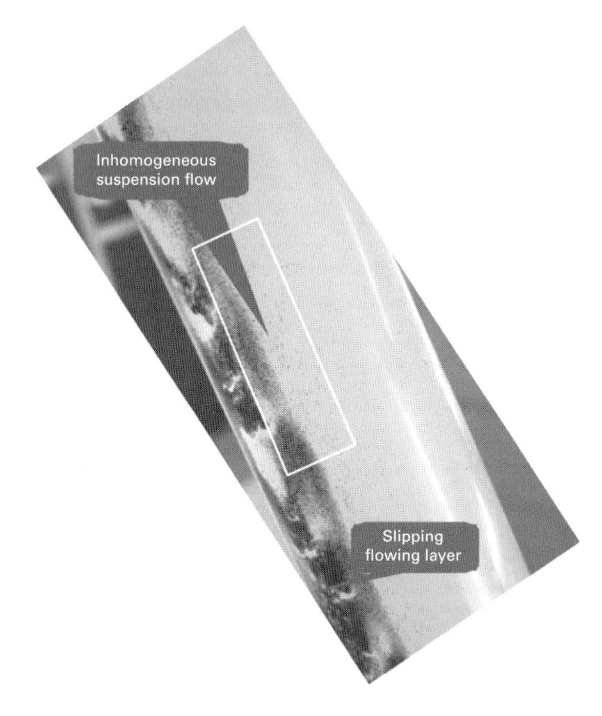

Figure 5-5 Picture of flowing of sand particles in tilting pipe at 45°.

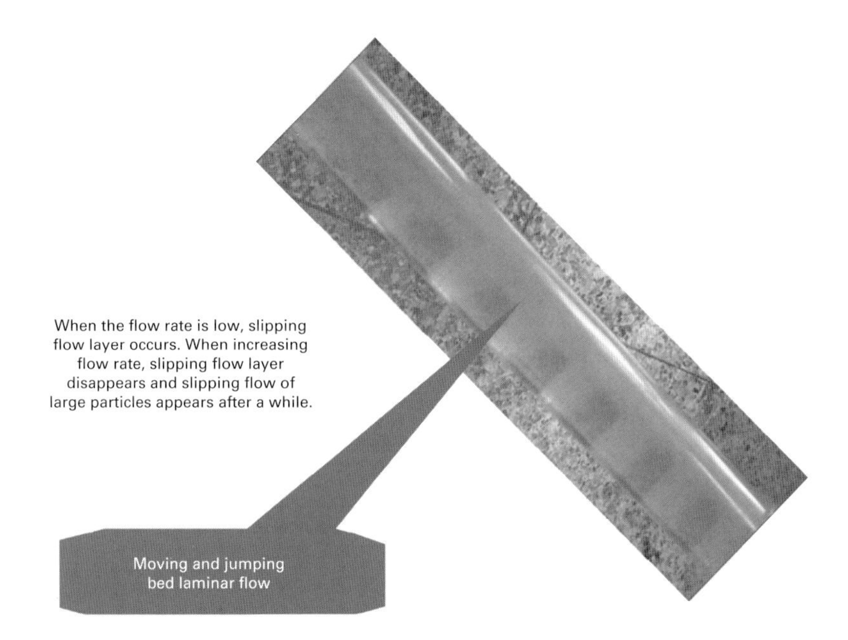

Figure 5-6 Picture of flowing of sand particles in tilting pipe at 60°.

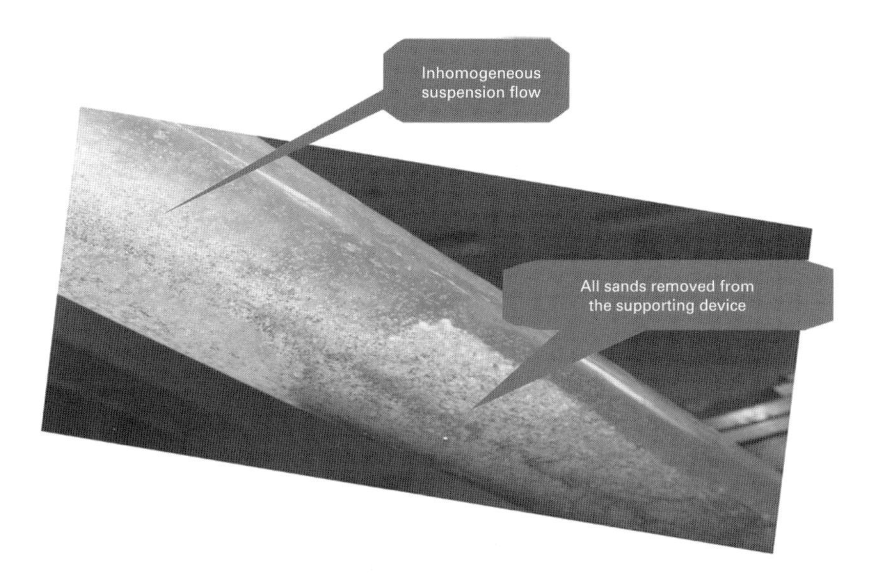

Figure 5-7 Picture of flowing of sand particles in tilting pipe at 75°.

Figure 5-8 Picture of flowing of sand particles in tilting pipe at 90°.

Table 5-3 Summary of Sand-Carrying Flowing Pattern Experiments

Dip Between Sand-Carrying Tube and Surface (°)	*0° (Horizontal Tube)*	*30°*	*45°*	*60°*	*75°*	*90° (Vertical Tube)*
Flowing pattern of the movement of sands	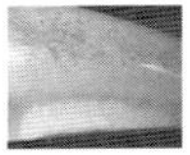	 	 			
Flowing pattern of the movement of sands	Jumping and moving bed, rolling forward gradually.	Moving and jumping bed laminar flowing. Obvious slipping forward of sand dune can be observed when the flow rate is low.	Inhomogeneous suspension flow and slipping laminar flow. Obvious upper flowing forward and lower slipping downward occurs when flow rate is low. Slipping laminar flow when increasing flow rate. Slipping laminar flow will be transferred to inhomogeneous suspension flow when increasing flow rate further.	Moving and jumping bed laminar flow. Small particles form slipping laminar flow with low flow rate; the slipping laminar flow disappears with the increase of flow rate, forming slipping laminar flow of big particles after a while.	Inhomogeneous suspension flow	Homogeneous suspension flow

1. When the sand-carrying pipe is placed vertically, sand particles move in the manner of uniform suspension flowing.
2. When the sand-carrying pipe is tilting, sand particles will move in the manners of nonuniform laminar flowing, jump and moving bed flowing. If the tilting angle is small, sand particles move mainly in the manner of laminar flowing; if the tilting angle of pipe is large, sand particles will move mainly in the manner of jumping and moving bed flowing.
3. For sand particles that transport in the manner of jumping and moving bed flowing in horizontal wellbore, some of them forming the main transportation mainly move in the manner of non-uniform laminar flowing and are carried out from the sand-carrying pipe.

5.1.3 Analysis of Experimental Data and Improvement of Formulas

5.1.3.1 Analysis of experimental data

According to the experimental data, the relationship between critical sand-carrying velocity in wellbore and the diameter of sand particles is shown in Figure 5-9, and a cubic polynomial can describe this relationship accurately, e.g. expressed by $u_c = a_1 d_p^3 - a_2 d_p^2 + a_3 d_p - a_4$ and the correlation coefficient R^2 is above 0.92. In this equation, the unit of diameter of sand particles d_p is mm and a_1, a_2, a_3, a_4 are coefficients related with well inclination. The range of a_1 is between 0.2 and 6.5, a_2 is between 0.2 and 9.0, a_3 is between 0.2 and 5.0, and a_4 is between –0.02 and 0.5.

5.1.3.2 Correction of formula

By comparing and analyzing experimental data and results of theoretical formula, the coefficient α in Equation (5-1) to Equation (5-6) can be achieved through the equation of $\alpha = 1.1 + Ae^{-Bd_p}$. A and B are constants that are related to well inclination and flowing pattern. The range of A is between 6 and 10, and the range of B is between 1300 and 6000.

5.1.3.3 Software for solving critical sand-carrying velocity

On the basis of combining sand-carrying experimental data, basic sand-carrying theory, sand production prediction of formation, fluid properties of multiphase effluent, and methods of calculating flowing velocity and flowing rate, we developed software of solving critical sand-carrying flowing velocity in wellbore which applies to various conditions like different fluid properties

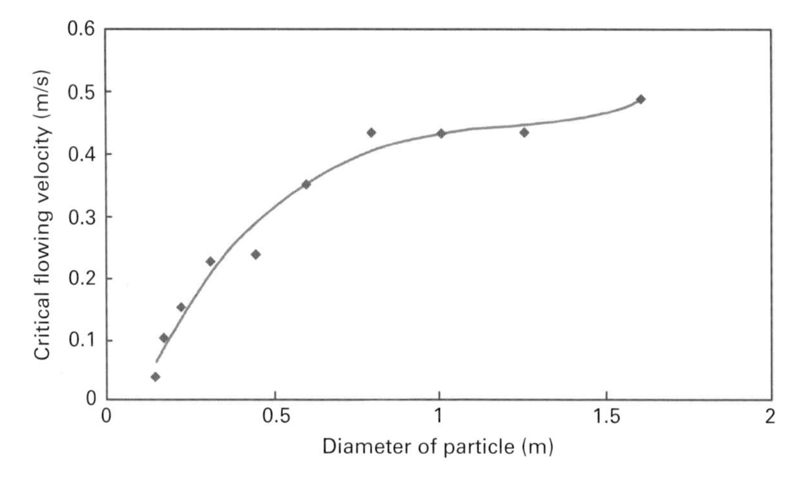

Figure 5-9 Relationship between critical sand-carrying velocity in wellbore and sand particle diameter under certain inclination of wells.

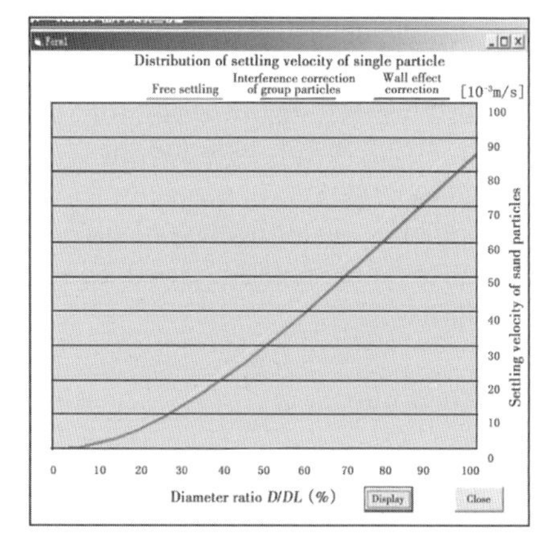

Figure 5-10 Interface of software to calculate critical sand-carrying velocity.

and flow rate. The interface of this software is shown as Figure 5-10. This software can calculate the parameters such as critical sand-carrying velocity and maximum sand-carrying concentration under certain conditions. Figure 5-11, Figure 5-12, and Figure 5-13 show the relationship plot considering critical sand-carrying velocity, well inclination, diameter of sand particles, and predicted sand concentration, where the viscosity of fluid-carrying sand is 120 mPa·s with three different pipe sizes.

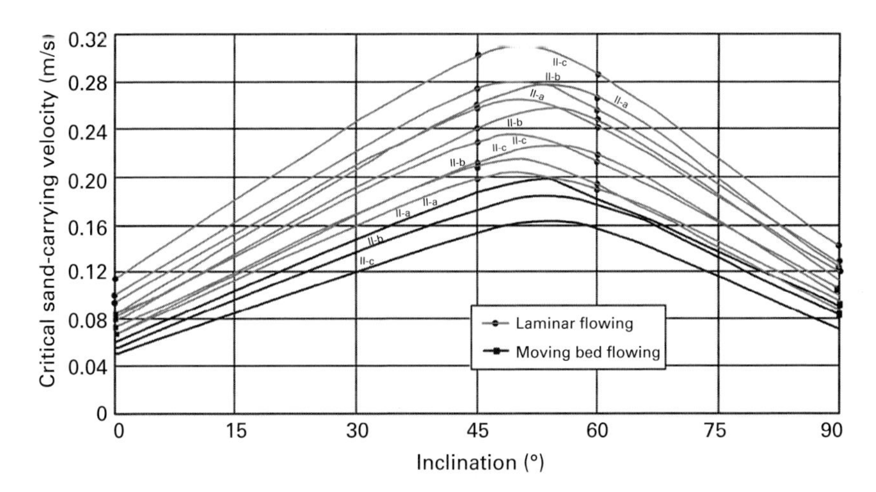

Figure 5-11 Relationship among critical sand-carrying velocity, well inclination, and sand concentration in tubing with inner diameter of 77.8 mm.

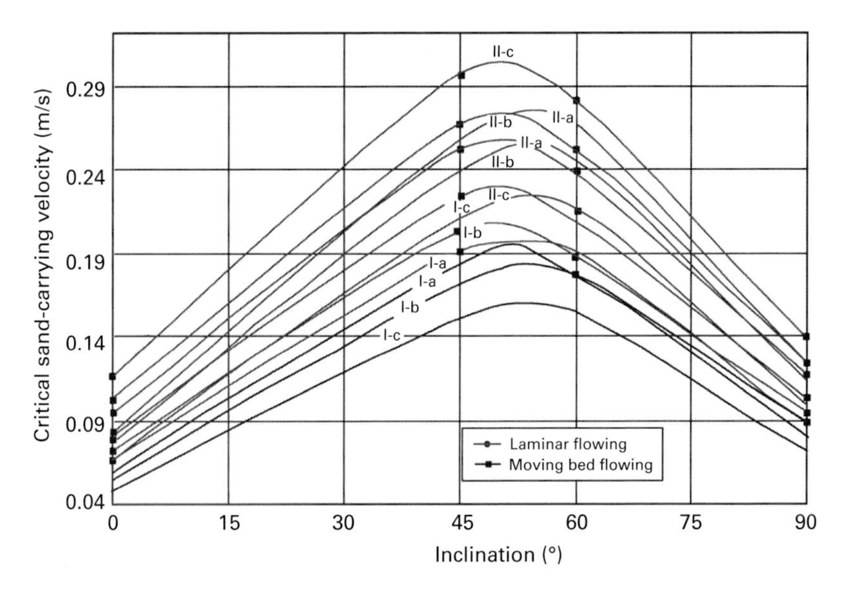

Figure 5-12 Relationship among critical sand-carrying velocity, well inclination, and sand concentration in tubing with inner diameter of 62 mm.

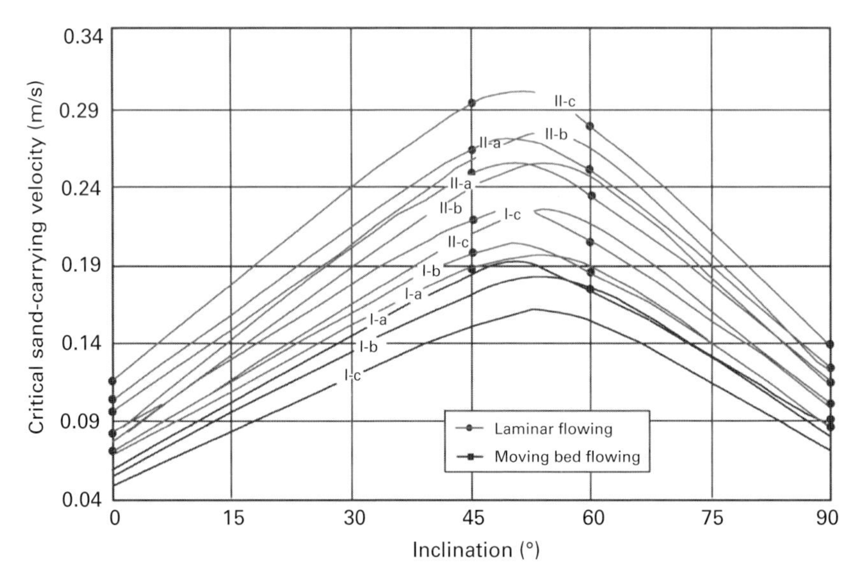

Figure 5-13 Relationship among critical sand-carrying velocity, well inclination, and sand concentration in tubing with inner diameter of 51.8 mm.

5.2 Techniques of Artificial Lift and Production Optimization When Producing with Sands

5.2.1 Artificial Lift and Supporting Techniques

The rate of well effluent that can carry produced sand in time and reasonable artificial lift techniques are an important step to ensure normal and efficient production of wells, also a key factor that determines whether the technique of producing with limited sand is successful. These techniques will ensure the produced sand is brought to surface in time, preventing the sand particles from settling in the wellbore and blocking wellbore, extending the life of downhole production devices and tools, reducing the frequency of repairing pumps, flushing deposited sand and operating costs, and achieving the goal of maximizing economical profit. Therefore, studies of sand-carrying fluid in wellbore and artificial lift techniques are very important and valuable.

5.2.1.1 Techniques of artificial lift

The suitable artificial lift techniques for sand production management are mainly electrical submersible progressive cavity pump (PCP), surface rod driving PCP, sand-proof electric submersible pump (ESP), and so on. All these equipment have their own features and application conditions.

Electrical centrifugal pumps have the features of wide ranges of flowing rate, high discharge head, and motor power and are widely used in oilfields for enhancing production rate. But the pump is also with the weak capacity of carrying viscous fluid, low pumping efficiency, and high energy consumption, which is why it is not easily adaptable to heavy oilfields with low or medium water cut. Currently, the ESP can be categorized into sand-resistant pumps and non–sand-resistant pumps. For non–sand-resistant pumps, the sand concentration of producing liquid should not be over 0.3%; for sand-resistant pump, sand concentration in producing liquid should be lower than 5%.

Progressive cavity pumps feature simple structure, small size, and light weight. Such situations like the cases of pump stuck, gas lock, blocking by sand or wax, or emulsion will not occur to these pumps. Progressive cavity pumps are with characters of capacity of carrying produced sand and viscous fluids, and high pump efficiency. Energy consumption is only one third of ESP for the same situation, so PCP is particularly suiTable for producing heavy oil or crude with produced sand. Progressive cavity pumps manufactured by MAPE of France can work pretty well when sand concentration is reaching 60%. At present, the practical maximum production rate of PCP is less than 200 m^3/d, and pump head is less than 2000 m. Progressive cavity pumps used in well production can be sorted into two types: surface driving PCP and electric submersible PCP.

Surface driving PCP producing system is mainly composed of three parts, namely, surface driving device, surface control system, and downhole pump and production string. PCP is pretty suitable for vertical wells, and its application in onshore vertical wells is very successful. PCP is the main technique for producing oil wells with sand, heavy oil wells and polymer flooding wells. When being applied in directional wells, due to the constrains of well inclination and dogleg severity, the pumping rod can run in a long term if it works with a slow speed, but the rod will break down after running for 2 to 4 months with high speed, resulting in short life term and the difficulty of increasing production rate. Researches and pilot testing of PCP in recent years found that it is very difficult to extend the life of PCP even high grade solid rods and high strength hollow rods are used. PCP is mainly used in directional wells with low production rate and deep kick off point, with the pump installed in vertical interval. In October 2007, CNOOC launched a pilot project of applying coiled rod driven PCP and achieved good results in this domain. The detailed information of this successful application are as follows: oil viscosity in formation is 2,056 mPa·s, the inclination where pump is installed is 68°, maximum inclination is 73°, the submersible depth is only 100m and production rate is $20m^3/d$.

The system worked continuously for 700 days until August 2009. By the beginning of 2009, the trials were expanded and the technique of coiled rod driving PCP was used in almost 20 wells.

From a certain perspective, electric submersible PCP combines the advantages of PCP and ESP, e.g. the combination of mature electric submersible motors and high efficiency progressive cavity pump. The downhole system of electric submersible PCP is composed of electric submersible motor, gear speed reducer, protector, flexible shaft, submersible PCP and so on. Production string accessories like oil drainage valves, pressure detection valves, and surface system are the same as ESP. The factors constraining the life of electrical submersible PCP are the speed reducer and protector. The technique of electric submersible PCP is the main method and objective of developing offshore oilfields producing with limited sand, which only consumes one third energy of ESP with the same flowing capacity (i.e. pumping rate, pump head). The life of electric submersible PCP manufactured by factories outside China is almost the same as that of ESP. The electric submersible PCP whose pumping rate is less than 200 m^3/d made by Chinese factories can also be applied in fields, and the life term is more than 400 days. For the pumps with pumping rate more than 200 m^3/d, the researches and practices in the recent years showed that the life term is less than 200 days and further researches are needed.

On the basis of the characters of surface well locations of offshore oilfield, the technical features of artificial lift methods, well type and well trajectory, well productivity, flowing pressure of bottom hole, the changes of water cut and the requirements of environment, the following criteria are used to select suitable artificial lifting techniques.

1. Screening criteria for selecting:
   ① Production rate is more than 200 m^3/d
   ② Sand concentration is less than 3‰
   ③ Water cut is higher than the phase inversion point of emulsion of oil
2. Principles of pumping rate design of electric submersible PCP:
   ① Production rate is between 100 m^3/d and 200 m^3/d
   ② Sand concentration is more than 3‰
   ③ Water cut is less than the phase inversion point of emulsion of oil
   ④ The depth of kickoff point is less than 300 m, and flowing fluid level is deep
3. Principles of pumping rate design of surface driving electric submersible PCP:
   ① Production rate is less than 100 m^3/d
   ② Sand concentration is more than 3‰

③ Water cut is less than the phase inversion point of emulsion of oil
④ The depth of kickoff point is more than 300m, and the flowing fluid level is high

When designing and choosing parameters of PCP and ESP, besides considering oil viscosity, density and GOR, sand concentration should be taken into account (e.g. the influence of sand concentration change on effluent density, friction along flow path, wear of pump and so on). Generally, the arithmetic weighting average of discharge head, motor power based on sand concentration by weight is conducted; at the same time, a certain surplus of pump head and motor power is reserved for meeting the requirements during life term of pumps, and the values are enlarged by 30% to 50%.

Currently, up to 800 wells are using ESP in Bohai Bay offshore oilfields, 20 wells using electric submersible pumps, 6 wells using surface driving PCP. The production strings used are shown in Figure 5-14 to Figure 5-17, respectively.

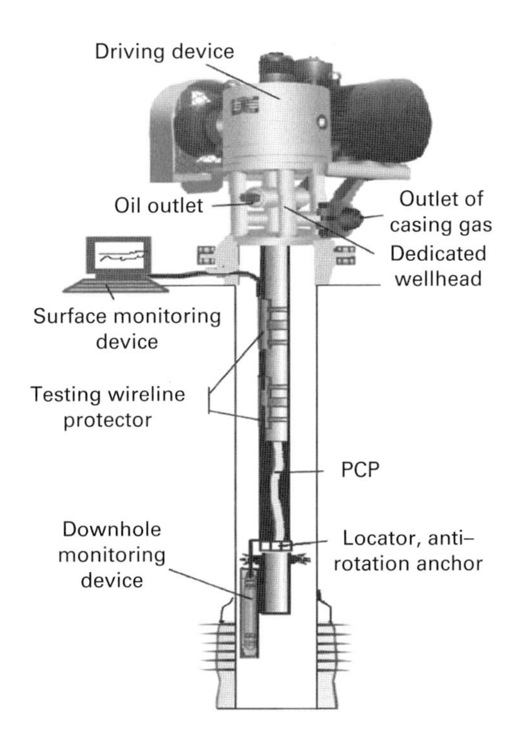

Figure 5-14 Diagram of production string and control configuration of surface driving PCP.

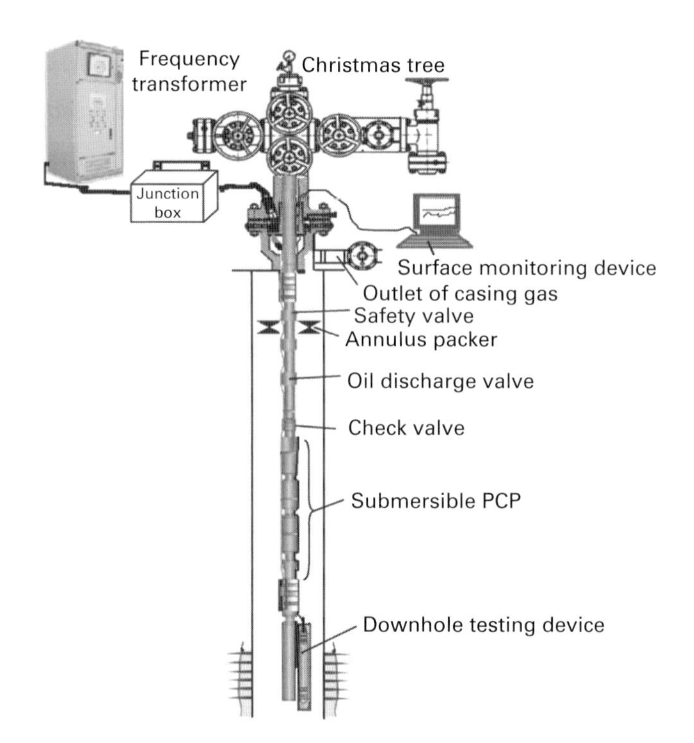

Figure 5-15 Diagram of production string and control configuration of electric submersible PCP.

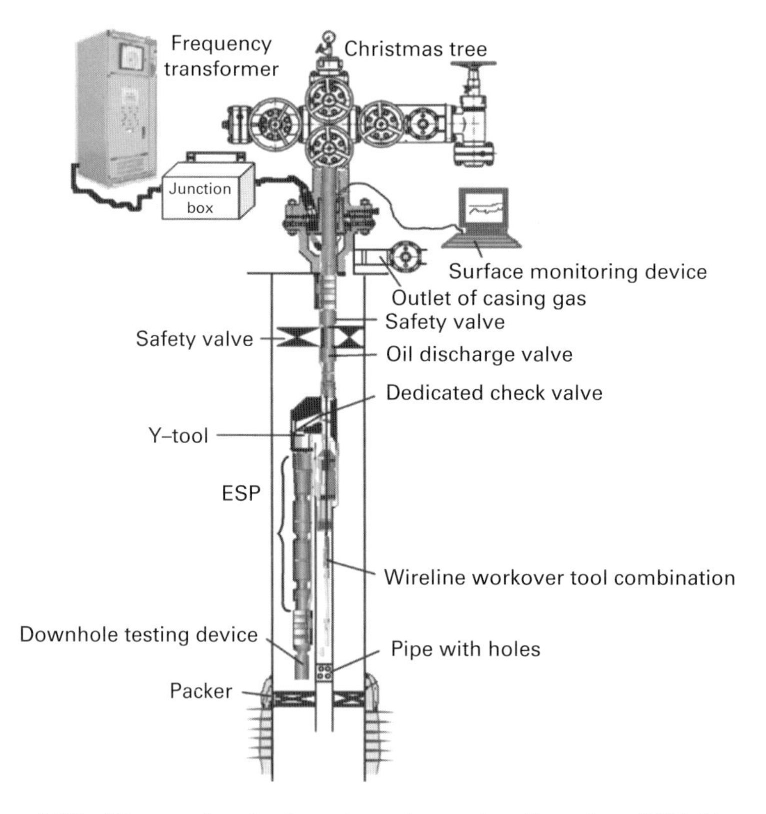

Figure 5-16 Diagram of production string and control configuration of ESP (1).

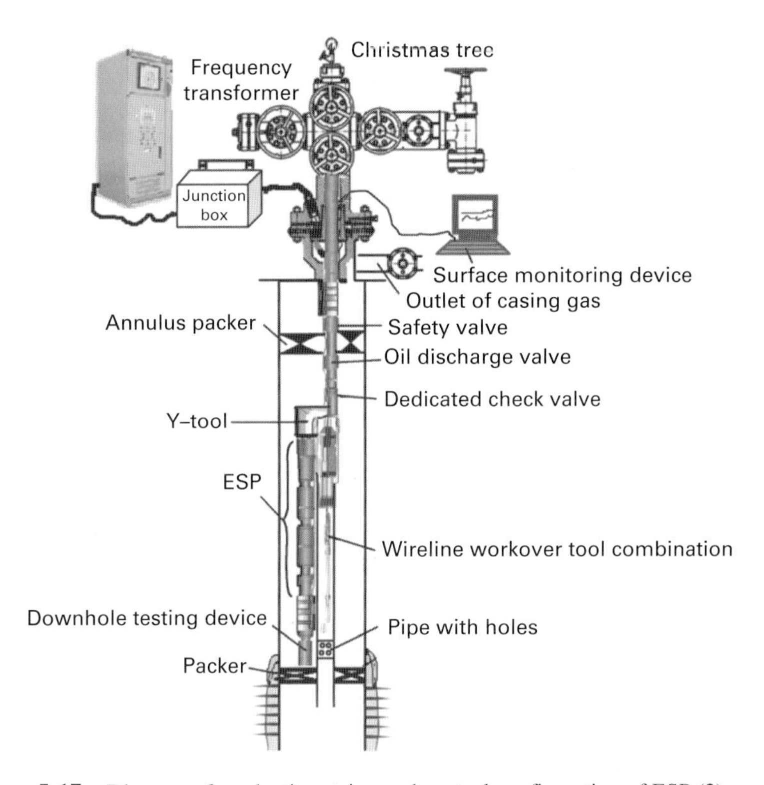

Figure 5-17 Diagram of production string and control configuration of ESP (2).

5.2.1.2 Supporting techniques

1. Techniques of monitoring sand production
   Oil well surveillance includes monitoring production rate, water cut, sand production, and downhole temperature and pressure. Production rate and water cut are commonly monitored by conventional methods, and downhole temperature and pressure are monitored using ESP surveillance instrument, PSI, PHD, and capillary tube and so on. Most of these monitoring tools can be manufactured in China and thus will not be discussed further.

   It is essential to monitor sand concentration in the produced liquid when allowing limited sand production during development. At present, the methods for monitoring sand concentration include laboratory testing, acoustic measurement, ultrasonic measurement, ray measurement, photoelectric method, infrared method, laser method and so on.

   Because signals of ultrasound, infrared, photoelectric are hard to penetrate multiphase mixture of oil, gas, water, and sand, these techniques are not

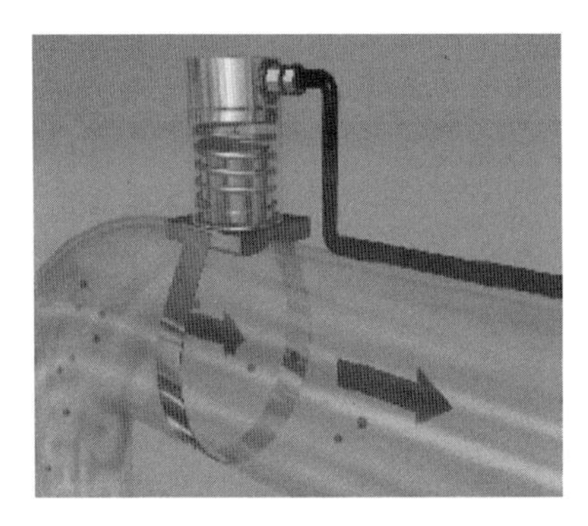

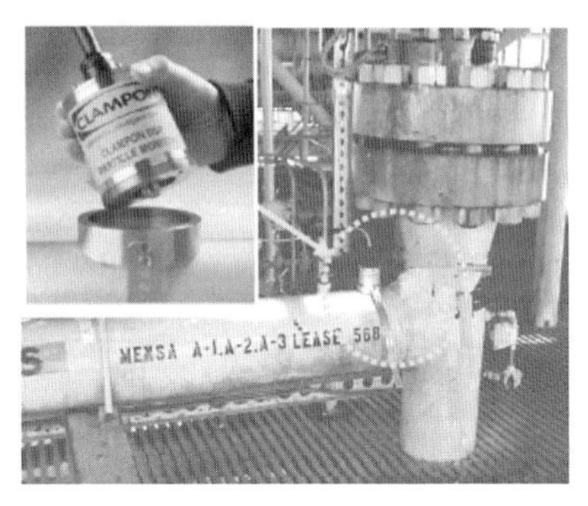

Figure 5-18 Diagram of external sand concentration detector installation.

implemented in this research. In addition, the method of ray measurement is not further researched either because it needs three different radiation sources to measure density of multiphase fluids including sands to identify sand concentration, which is suitable for wells with massive sand production. Also, the application of ray measurement method is not considered due to its complicated system, multiple radiation sources, and complex measuring circuit.

Acoustic method to identify sand concentration is based on measuring acoustic signal of friction between sand particles and pipe wall, which is not harmful to operators and environment. This technique can achieve continuous monitoring, and it occupies less space. So, acoustic method is regarded as the best measuring method of monitoring sand concentration. The detector used in this technique includes intrusive detectors and external detectors. The intrusive detector is to insert probe vertically into the wall of pipes to receive the collision signal of sand particles against probe, and its disadvantage is that the flowing sand particles will wear probe easily. External detector is to install the acoustic sensor in the outer wall of 90° elbow in the pipe. The advantages of this technique are easy installation and less damage, so external acoustic surveillance device is used to monitor sand production. ClampOn 2000/2100 particle/pigging detector manufactured by ClampOn Corporation is used in Penglai 19-3 oilfield (Figure 5-18). This device is used to achieve sand concentration per barrel or per ton (expressed in kilograms) by measuring the total volume and velocity of multiphase flowing. Figure 5-19 shows the well performance of a certain well, and the sand concentration in the plot is converted to per mille concentration (‰) by volume by using density transfer.

Artificial measuring approach is a conventional method of monitoring sand concentration, which consumes lots of time and workload, and the measuring result only represents the status of a certain period. It cannot

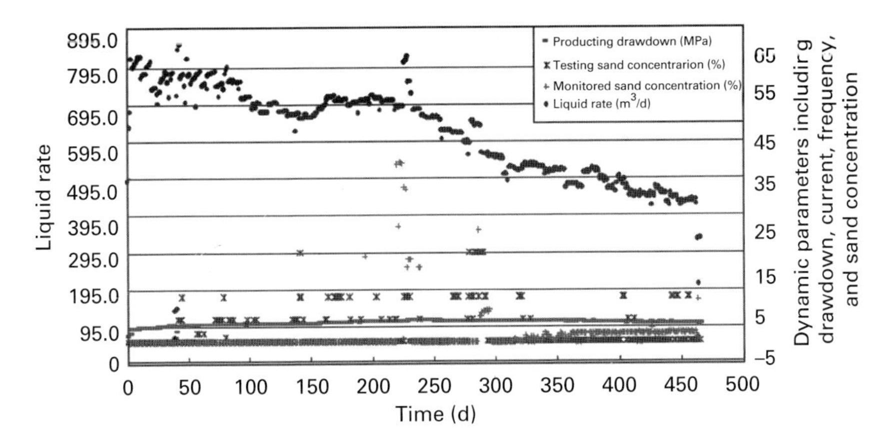

Figure 5-19 Well performance of a sanding well in a certain oilfield.

realize continuous monitoring and poorly represents the true sand concentration, so it is only used for calibration, verification, and as a replacement of continuous monitoring devices when they fail.

Figure 5-19 shows that there are some difference between artificial measuring approach and continuous monitoring method. The major reason is that artificial analysis result only represents samples at a certain time, while the continuous method reflects the average value of the 24 hours in a whole day.

The well shown in Figure 5-19 is a pilot sanding well, and the data are collected from startup of production to 460 days after. The frequency of ESP of this well was 30 to 50 Hz in the first 160 days, the volume of sand production was medium and production rate of this well was maintained at 700 to 800 m^3/d. In order to test the maximum volume of sand production, the frequency of ESP was improved from 55 to 60 Hz gradually from the 161th day. Sanding peak occurred two months later after this and was maintained for 2 more months. Sanding volume then decreased sharply and sand concentration remained at 0.5% to 1.5%.

2. Techniques of cleaning up plugging formation
After producing for a long term with sand, the mixture of some sand particles, clay, scale, wax, resin and paraffin will settle in the slots of screens, gravel packing zone, perforated tunnels and the zone surrounding wellbore due to the blocking effect of screens, which results in the blockage of flowing paths and decreasing well PI. This kind of mixture is hard to clean up by conventional acid or organic solvents. Through years of researches and repeated applications in fields, the technique of cleaning up

formation blockage without moving production string and composite multifunctional cleaning blockage formula are achieved, which are developed to commonly used stimulation measures and prove to be profitable.

The main performance indicators of the composite multifunctional chemical of cleaning blockage are as follows: the ratio of cleaning organic sediment is not less than 95% (60°C, 2 h); ration of scale dissolution is not less than 95% (60°C, 2 h); ration of core dissolution is not less than 15% (60°C, 4 h); permeability of core increases by 1 to 5 times; ratio of corrosion is 2 ~ 3 g/(m^2·h) (90°C, 4 h); good compatibility with formation.

3. Techniques of prevention of sand deposition and well flushing
More or less sand particles will settle down in the wellbore after shutting down for a period of time, especially in the stage of high water cut. When sand deposits in the outlet of pump, reopening of the well will be affected. When sand settles and accumulates in the bottom of wellbore, the PI of wells will be affected. Researches and practices find that, for wells with electric submersible PCP and surface driving PCP and ESP wells with high rate and low water cut, it is less likely that sand particles settle in wellbore above pump due to high viscosity of effluents of wells, and production string shown in Figure 5-14, Figure 5-15, and Figure 5-16 can be selected for this case. For the ESP wells with high water cut, sand particles will deposit in the wellbore after shutting down the pump due to gravity separation of mixed effluent with low viscosity, so production string shown in Figure 5-16 is used. When producing with ESP, the special check valve is closed and the well works normally; when shutting down the ESP, the special check valve will be opened and sand particles above pump will fall down to the wellbore below the pump to prevent sands from clogging the outlet of the pump and tubing.

As for flushing liquid, a series of non–pollution flushing chemicals are researched and applied like nondamaging foam flushing liquid, low damage flushing liquid, and so on. Low damage flushing liquid is mainly used in the stage of high formation pressure, and nondamaging foam flushing liquid is applied in the stage of low formation pressure. In general, foam cleaning chemical is implemented when the formation pressure coefficient is less than 0.80.

The flushing chemicals are with the following characters: small volume of liquid filtration loss, little damage of filtration on formation, strong ability of carrying sands and cleaning tubing and casing, good compatibility with treating chemicals of water and oil, and the agents meeting the requirements of corrosion ratio of industrial standards.

The application of nondamaging flushing chemicals reduces greatly the loss of flushing liquid and damage on the formation, shortens the recovery cycle of wells from 10 ~ 20 days to less than 5 days and increases effective production time of wells.

4. Production string
   For the case of oilfields in Bohai Bay, the types of production strings shown in Figure 5-14, Figure 5-15, Figure 5-16, and Figure 5-17 are mainly used during development.

5.2.2 Production Optimization

5.2.2.1 Principles of control and adjustment of production rate

The basic principles of controlling and adjusting production rate in artificial lift wells in offshore oilfields are as follows:

1. Control production drawdown in a reasonable range to allow the production of fine and silt sands and form stable sand arching, and prevent screens from damaging.
2. Maintain a reasonable production rate and release the potential of oil wells.
3. Keep the mechanical devices in a reasonable scope, especially under the optimal conditions if possible, and running the equipment in low energy consumption status.

5.2.2.2 Techniques of controlling and adjusting production rate

In order to achieve the above-mentioned objectives, the integration of various techniques such as adjusting production rate using variable frequency and choke, real-time closed-loop pressure control of variable frequency unit, real-time monitoring of downhole pressure, measurement of fluid level, continuous pressure measurement using capillary tube, PHD continuous pressure measurement and PSI continuous pressure measurement is implemented. Production drawdown and the operating point of the pump are controlled in a reasonable range, as shown in Figure 5-20.

For the wells that were put on production before 1996, the major techniques used in the pumping wells were to control by working frequency drive, to adjust production rate by choke, fluid level measurement, capillary continuous pressure measurement, PHD continuous pressure measurement, and PSI continuous pressure measurement. The approach of "one variable, multiple control methods" was applied in fields. Reservoir engineers and production engineers analyze the reasonableness of production drawdown and the pump

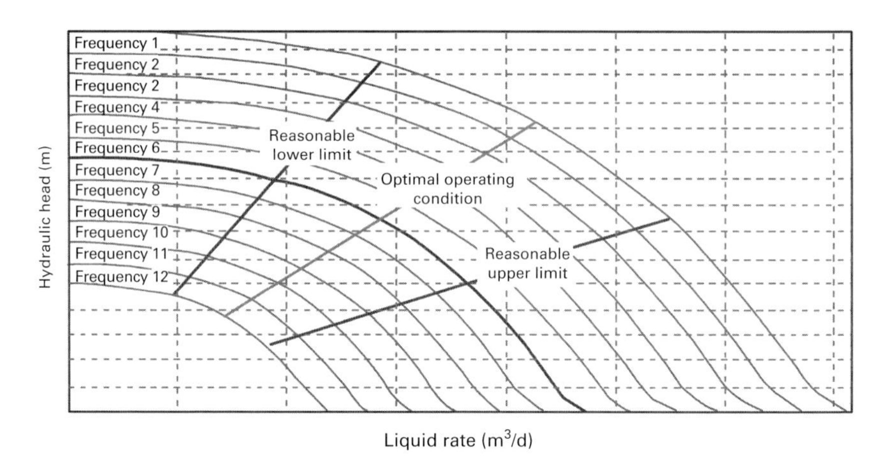

Figure 5-20 Controlling operating point of the pump and oil well by variable frequency unit and choke.

together on the basis of acquired data such as downhole pressure, well drawdown and operating parameters of artificial lift units, then identify the reasonable size of choke and give the data to operating engineers to execute. When the adjustment of the choke cannot meet the requirements of production rate adjustment, operating parameters of the EPS system will be adjusted to meet the requirement of production rate.

For the wells put on production from 1996 to 2005, well counts using capillary testing increased on the basis of previously used measuring techniques. Some of the wells use permanent pressure gauges, and the approach of "one variable, multiple control methods" of variable frequency unit is commonly used in new oilfields. Soft start of variable frequency unit is realized and production rate of some wells can be adjusted using variable frequency. When putting a well on production or restoring shut-in wells, a low frequency of variable frequency unit is applied to start a well. When the frequency is increased to 50 Hz gradually, the system will switch to working frequency driving control system automatically to prevent massive sand production from bottom hole pressure surge.

For the new oilfields put on production after 2005, the surface control system of artificial lift wells is configured with variable frequency drive by the model of "one on one", and the method of starting with low frequency and adjusting production rate by changing frequency is applied. One third of the wells use ESP real-time monitoring, one third of them use permanent pressure real-time monitoring technique, and some of the wells apply sand concentration surface monitoring technique. The systematic configuration is a good base for dynamic tracing and production adjustment. The techniques can maintain well drawdown in a reasonable range in time and prevent the failure of sand excluding system due to great drawdown. Also, the production rate can be controlled conveniently by combining with adjusting production rate with choke in the life term of ESP instead of the workover of changing pump and so on. Therefore, the operating point of ESP will be maintained in a reasonable and efficient range to both extend the life of pump system and save energy. On the basis of recording and analyzing the surface sand concentration, downhole pressure, production rate and the frequency of pump simultaneously, the reasonable empirical values achieved from wells can be applied in the new offset wells in the same field where sand concentration monitoring device and downhole pressure gauge are not installed.

In order to realize automatic management of oilfield production, the experiment of controlling well production rate using the pressure closed-loop technique is carried out. In other words, we input the important parameters of wells such as sand concentration, downhole pressure, temperature, production rate, drawdown, signals of operating status of ESP, and so on into variable frequency drive control system as appraisal data of adjusting operating frequency of variable frequency drive, so the fully automatic adjustment of production rate in fields can be achieved.

The detailed processes of opening wells using variable frequency drive are as follows:

1. After controlling the quality of variable frequency drive of ESP and surface system of downhole gauges, record the data of downhole pressure gauge like static pressure, and observe casing pressure and tubing pressure.
2. Open subsurface control downhole safety valve, wax cleaning valve, production wing valve of Xmas tree, surface safety valve, valve of choke and main valve subsequently, release tubing pressure, and let water, oil, and gas in.
3. Open four-way wing valve of tubing, release casing pressure and let oil, water and gas in.
4. Double check the wellhead valves to make sure the openness of valves to flowline, and record pressure data from gauges.
5. Set the starting frequency of variable frequency drive to 40 Hz, estimate wellhead pressure when the variable frequency drive is running forward at the frequency of 30 Hz on the basis of rated parameters of ESP and well pressure.
6. Start variable frequency drive and ESP, close production wing valve, build the pressure, and identify the turning direction of motor.

7. Open production wing valve after confirming the turning of motor is right, and maintain the production for 2 to 4 hours at the frequency of 30 Hz. The production drawdown is controlled at 60% of proposed value of development plan. At the same time, monitor sand concentration of produced liquid, downhole pressure, fluid level, and check the productivity index. If sand production occurs, reduce choke size to decrease production drawdown; if no sand production occurs, then go to the next step.
8. Increase the frequency to 35 Hz, and repeat the operations in step (7).
9. If no sand production occurs while increasing frequency, improve the frequency by 5 Hz per time or adjust to the suggested production drawdown or production rate.

5.3 Surface Treatment of Produced Crude with Sand Particles

It is essential for petroleum processing and refinery to clean up sands from oil and to remove oil from produced sands. To remove sands from crude oil is to reduce the abrasion, damage, clogging of sand particles to equipment of oil and gas treatment, storage and transportation and refinery, containers and pipelines, and the risk of safety issues, treatment costs, and environmental problems due to the damage of flowlines. The objective of cleaning up oil from produced sands is to avoid environmental pollution due to the treated produced sands. For light oil reservoirs, to remove oil from produced sands and to remove sands from crude oil are pretty easy, but it is very hard to do so for heavy oilfields, especially in the stage of low water cut.

5.3.1 Characteristic of Oilfields in Bohai Bay and Requirements of Treatment

5.3.1.1 Characteristic of oilfields in Bohai Bay

According to the developed and to-be-developed offshore heavy oilfields in Bohai Bay, the following features can be summarized:

1. Heavy oil reserves account for a large portion of total reserves in this area, and there are lots of categories like conventional heavy oil, medium heavy oil, and extra heavy oil. The viscosity in some reservoirs is sensitive to temperature; while that of other reservoirs is sensitive to solution gas.
2. Most of the wells in this area are with low water cut (less than 5%) in the early stage after putting on production (some even without water).

The wells will go through stages including low water cut, medium water cut and high water cut when developing with water injection.

3. Crude oil is with high concentration of resin and paraffin, generally more than 30%. When water breaks through, serious emulsion will easily occur. To produce with ESP makes the emulsion even worse, and the more stages of ESP, the more serious emulsifying. The increasing ratio of the viscosity of emulsion will be larger if the emulsion drops are smaller.
4. The water cut of phase inversion point of oil and water emulsion is 40% to 50%. When water cut is less than that of phase inversion point, the type of emulsion is W/O (emulsion) and the viscosity of oil is 5 to 7 times of that of oil without water. When water cut is higher than that of phase inversion point, the type of emulsion is O/W (reverse emulsion) and the viscosity of oil will decrease sharply.
5. High water cut and great accumulation of producing water are beneficial for separating water from oil and removing oil from water.
6. The unconsolidated formation is potentially easy to produce sands and cementation is with high clay content. The concentration of fine silt sands and movable sands varies from fields to fields. The sedimentary environment of most of the reservoirs belongs to fluvial facies. The distribution of sand particle sizes is quite different in horizon, and there are many pays vertically. The intervals with oil are large, ranging from tens of meters to hundreds of meters, even thousands of meters, and the size of particles is very different. So, to conduct sand exclusion in these reservoirs is very hard, and any error in execution will possibly result in the failure of sand exclusion.
7. Efficient high-speed development of these fields is required due to the limit of surface facilities of offshore oilfields, so the production rate of wells is high.
8. It is inconvenient to develop these fields because they are far from mainland and the transportation cost is high.
9. The ocean is one of the most important fishing sites, where ecosystem is very complex, so stringent environment protection is required.

5.3.1.2 Requirements of oil and gas treatment and disposal

1. Bohai Bay is an important fishing area and a level one national pollution discharging area. The waste generated during production must comply with the relevant national standards. Before May 1, 2009, *Industrial Oily Waste Water Discharging Standards of Offshore Petroleum Development (GB 4914-1985)* was implemented. After May 1, 2009, new national standard *Pollutant Emission Concentration Limit* of *Offshore Petroleum Exploration and Development* must be executed (as shown in Table 5-4).

Table 5-4 Discharging Standards of Offshore Petroleum Production

Item	*Grade*	*Concentration Limit (mg/L)*			
		Standards of Year 2008		*Standards of Year 1985*	
Petroleum category	Regional	Permitted value at one time	Monthly average	Permitted value at one time	Monthly average
	Grade 1	≤30	≤20	45	30
	Grade 2	≤45	≤30	75	50
	Grade 3	≤65	≤45		
Production garbage	No discharge and disposal into the sea			Not specified	

2. Because the surface treatment devices and flowlines for disposing waste water are costly, the treatment equipment should be with small volume, short flowline, compact structure and low energy consumption provided that the treatment quality is guaranteed.
3. With the high-speed development of techniques such as allowing limited sand production and CHOPS in heavy oil reservoirs, it is likely to produce some solid waste like sands and silt which is not allowed to discharge to the sea, so harmless disposal or reuse should be conducted for these produced materials. If the waste is transported to land for disposal, *Pollutant Control Standards of Agricultural Sludge* (GB 4284-1984) should be met, and oil concentration in sands must be less than 3‰ (percentile by weight).

5.3.2 Principles of Treatment

According to the properties of well effluent, the methods of cleaning up sands from crude oil containing sand particles are categorized as follows: (1) removing silts and sands by gravity; (2) removing sands by water washing and settling; (3) cleaning up sands by chemicals, water washing, and settling; and (4) removing sands by mixing light oil with crude oil downhole, mixing hot water at wellhead, flushing and discharging, and settling in processing pond.

The techniques of processing and disposing oily sands are as follows: (1) combustion method, by which oily sand is burned in situ or after being collected, then is discharged or for other use; and (2) hydrocyclone separation method, by which the sands with water and oil are separated hydrocyclone; the treated sands are very clean after separation and sand concentration after treatment is less than 0.5%.

Most of the oily waste water after treatment is injected into the pay zone to flood oil. If the water cannot be injected into reservoir for flooding, it will be injected into the state-allowed water formation. Some waste water will be discharged into the sea after treatment as long as the waste water has no formation to reinject.

The above-mentioned techniques and methods of separation of liquids containing sands, oil, and water follow the principles of Stokes.

For the treatment of fluids containing sands with high water cut or low viscosity, basically, the separation of fluid with sands and oil/water follow the principles of Stokes as well is conducted based on gravity difference. The design of container uses Equation (5-7) to calculate free settling velocity of sand particles in fluid.

$$\upsilon_{\mathrm{f}} = \frac{g\Delta\rho d^2}{18\mu} \tag{5-7}$$

Where g is gravitational acceleration, m/s^2; $\Delta\rho$ is density differential between sands and mixed fluid (or density differential between oil and water), kg/m^3; d is diameter of dispersed particles of sands, oil or water, m; μ is viscosity of well effluent mixture, mPa·s.

But for the oilfields with medium or low water cut, high content of fine siltstone and high production rate, the size of separating equipment will be larger inevitably if only relying on gravity, which does not meet the principles of developing offshore oilfields efficiently and economically. In order to reduce the size of containers and equipment and save space, the technique of hydrocyclone is used to accelerate the process of de-sanding and increase the separating speed of particles based on centrifugal force and settlement. The moving speed of sand particles with centrifugal force can be calculated by the following formula:

$$\upsilon_{\mathrm{f}} = \frac{\pi^2 \Delta\rho n x^2 d^2}{2025\mu} \tag{5-8}$$

Where n is centrifugal rotating speed, r/min; x is distance of particles from the central point, m.

Also, the emulsion in the stage of low water cut is W/O (water in oil) and emulsified particles are significantly dispersed and very fine due to the influence of downhole artificial lift equipment.

Therefore, water is added to accelerate the accumulation, phase inversion of oil/water emulsion, the viscosity of the mixture of oil and water, and the thickness of emulsion. The principle of accumulation and phase inversion is also following the formula of concentration distribution:

$$\begin{aligned} c &= c_0 \exp\left[\frac{-mg\left(1-\rho_o/\rho_w\right)h}{kT}\right] \\ &= c_0 \exp\left[\frac{V\Delta\rho_{ow}h}{kT}\right] \end{aligned} \tag{5-9}$$

Where c is concentration of dispersed phase at a certain height, %; c_0 is initial concentration of dispersed phase, %; ρ_o is density of oil phase, kg/m^3; ρ_w is density of water phase, kg/m^3; h is thickness of emulsion, m; m is mass of sands, kg; k is Boltzmann constant; T is temperature, K; V is volume of continuous phase; $\Delta\rho_{ow}$ is density differential between oil and water, kg/m^3.

5.3.3 Examples of Surface Treatment of Fluid with Sands

Techniques and methods of high quality sand control are used in the production of offshore oilfields in the past twenty years, so the volume of sands produced from formation is controlled strictly under the acceptable range of surface production flowlines and processing equipment. In order to maintain the daily production, the measures and management system of cleaning equipment and flowline regularly are carried out.

Approaches of excluding sand and selecting and designing of sand exclusion tools face great challenges due to the development of multilayered, thick and great span pays. Massive sand production occurs because of the inevitable failure of sand exclusion and the damage of sand controlling tools. For these developing oilfields, sand settling and clogging in the measuring separator and production separator will occur because the surface desanding flowline is not considered and relevant desanding equipment is not installed; therefore, the normal production is influenced seriously.

On the basis of the past experience and lessons from practice, the features of development and requirement of offshore heavy oilfields, the simplified treatment devices of heavy oil mixed with sands and oily sands are illustrated as shown in Figure 5-21 and Figure 5-22. The techniques of remixing waste water to flowline and hydrocyclone separating as shown in Figure 5-23 are used to separate sands from crude oil completely as well as control water cut of oil within the standards of commercial crude oil and the oil concentration in sands within the onshore treating standards. The concentration of oil, sand, and solid of waste water after treatment will meet the standards of reinjection,

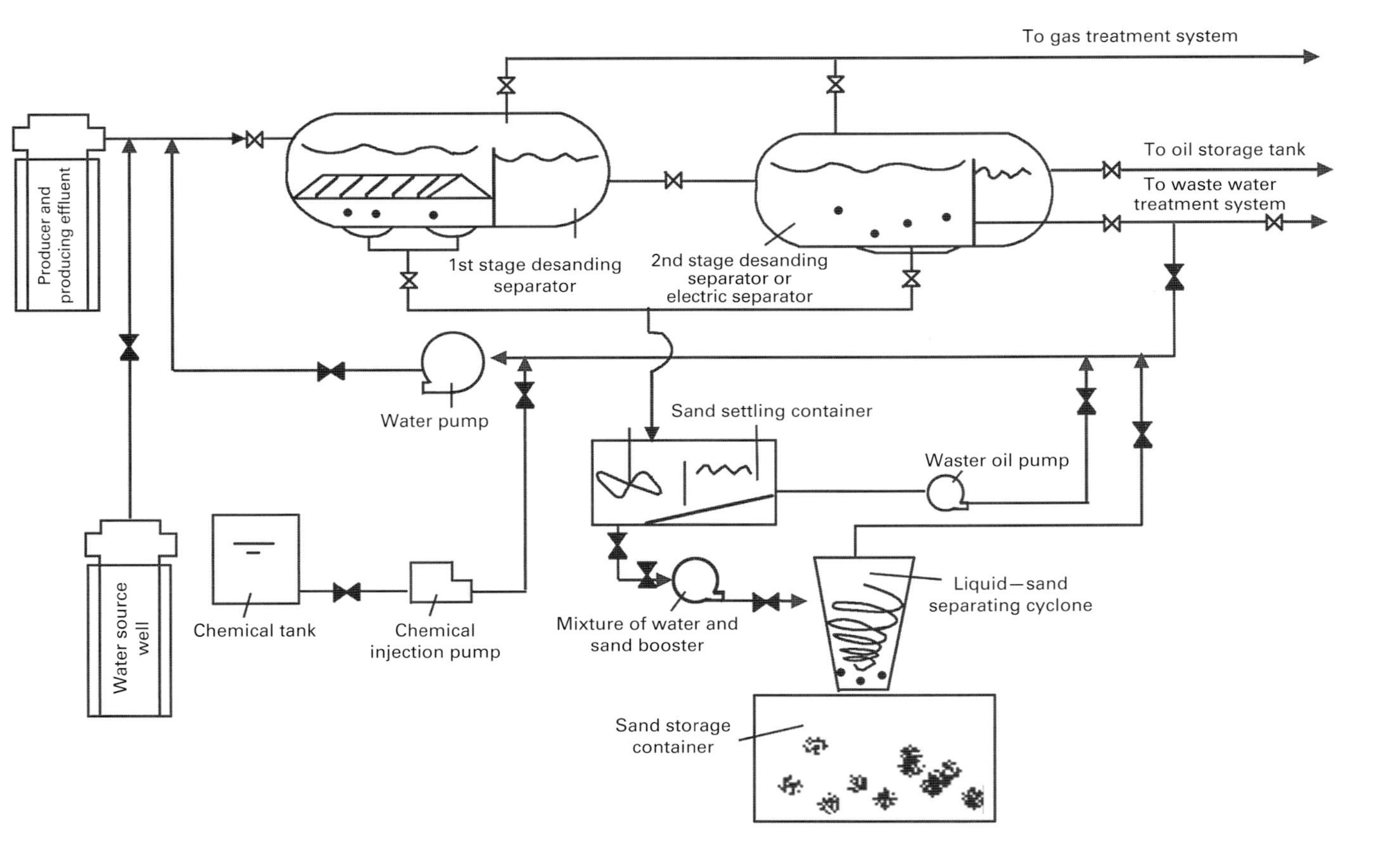

Figure 5-21 Desanding of oil and removing oil from produced sands in oilfields with low or medium water cut.

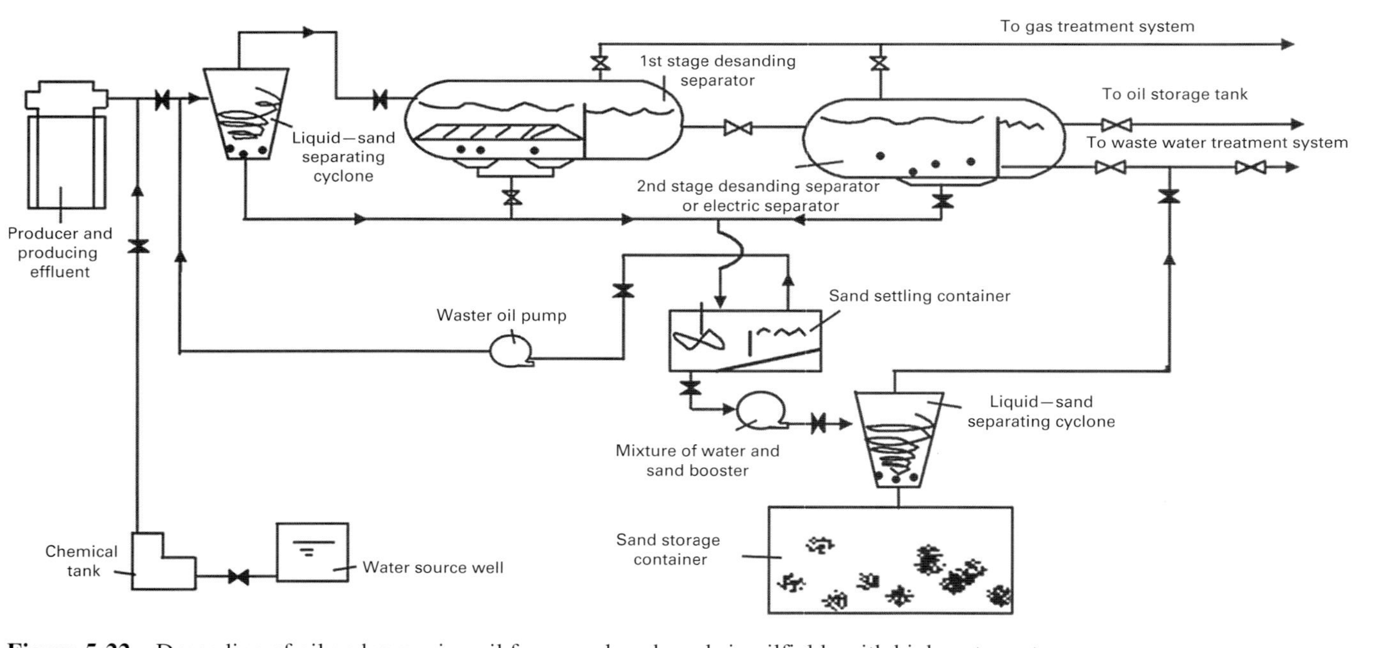

Figure 5-22 Desanding of oil and removing oil from produced sands in oilfields with high water cut.

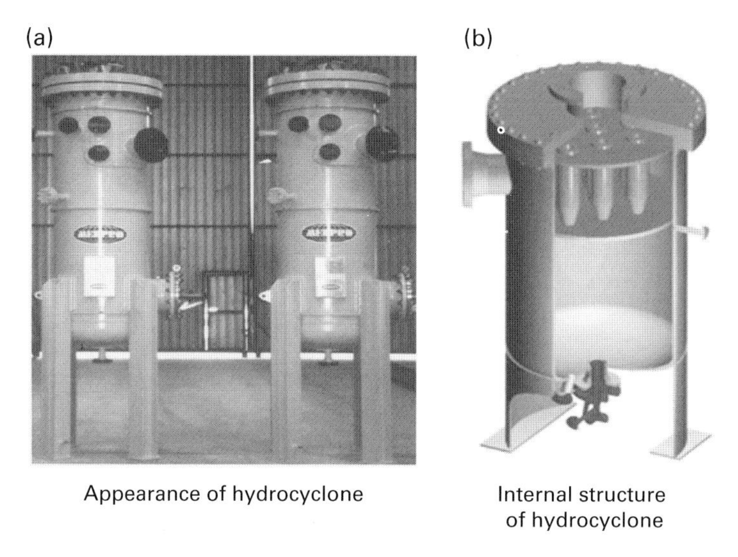

Figure 5-23 Liquid–sand separating hydrocyclone used in a certain oilfield.

and the requirements of discharging, and the techniques are applied successfully in practice. The flowing diagram shown in Figure 5-21 is mainly used in oilfields without water or with low or medium water cut (water cut is lower than phase inversion point of emulsion), and the flowing diagram shown in Figure 5-22 is mainly used in oilfields with high water cut.

Oil without water or low water cut produced from wells will mix with hot water produced from water source well and chemicals (e.g. demulsifier), and the mixture will flow into the first stage desanding separator to remove sand from oil. Then, oil after desanding and preliminary dewatering will flow into the second stage desanding separator (or electric separator) for the separation of oil and water, degassing or further desanding. Then the qualified oil after cleaning will be stored in tanks for sale (if degassing is not complete, one more stage for oil–gas separating can be added). Some of waste water separated from the second desanding separator will be cycled to the first stage desanding separator for diluting well effluent, and the left waste water after further treatment will be discharged to sea or used for other purposes as long as the waste water can meet desanding requirements of the first stage desanding separator when shutting down the water source well. The waste water with high sand concentration discharged from the first and second stage desanding separators flows to the sand settling container for deposition and complete separation of sands and oil. Then, waste water with high water concentration in settling container will be pumped to the first stage desanding separator for further disposal using waste oil pump, and oily mixture of water and sands at the bottom of container will be pressurized using lift pump to hydrocyclone for separation of sand and fluid. The fluid after separating will be discharged through the top line of hydrocyclone and flow back to the first stage desanding separator for further treatment after mixing with waste water cycling system. The sands after purifying by hydrocyclone will be discharged through bottom line of hydrocyclone, and will be drained to sea directly after quality controlling. The natural gas from the first stage and the second stage separators can be used for other purposes after the treatment of natural gas purifying and disposing system.

The features of the processing system are as follows:

1. Making full use of hot water to dilute and decreasing the viscosity of fluid to reducing the difficulty of desanding process; not sensitive to temperature; pretty suitable for oil with high viscosity or oil with less sensitivity to viscosity temperature; less energy consumption.
2. Greatly reducing the energy consumption using recycling mixed hot waste water and improving energy efficiency.
3. Small size and occupying less space due to separation of solids and fluid (sands and fluid) using hydrocyclone; better effect of removing oil from sands; the sands after treatment can be shipped to onshore for burying or other purposes.

6

Sand Production Management in Heavy Oilfields in Bohai Bay

The density of crude oil of conventional heavy oil reservoirs in Bohai Bay is at 0.82 to 1.02 g/cm^3, mainly between 0.95 and 0.98 g/cm^3, and the viscosity of oil in formation condition is between 50 and 450 mPa·s. On the basis of practical worldwide experience, the reservoirs which are suitable for sand production management are those with density of degassed oil being 0.934 to 1.007 g/cm^3, and viscosity being 600 to 16,000 mP·s. The fluid properties of most reservoirs in Bohai Bay are within these ranges, and most of them such as Luda, Caofeidian, and Bozhong oilfields are with high solution gas and are thus suitable for the technique of sand production management.

The rock strength of some blocks of heavy oil reservoirs in Bohai Bay is weak or even poor, so sand production will occur when applying techniques of conventional development and the cost will be increased and productivity index will be impaired if applying gravel packing sand exclusion. Implementation of the technique of sand production management requires that the formation rock is unconsolidated, with low clay content (usually less than 20%, preferably less than 10%) and without calcium cementation. The clay content of formations in Bohai Bay is within 5% to 35%, the rock is unconsolidated, and formations are with strong heterogeneity, all of which contribute to the success of the implementation of sand production management.

Sand Production Management for Unconsolidated Sandstone Reservoirs, First Edition.
Shouwei Zhou and Fujie Sun.

6.1 Core Technologies of Sand Production Management

Sand production management is a complicated and systematic technique that requires extensive researches and discussions. Screening the candidate oilfields, production interval and wells and conducting pilot field practice are necessary to achieve an integrated sanding management model before wide-scale implementation.

6.1.1 Studies of Formation Sanding Patterns

Researches of sanding conditions and sanding patterns are the basis of sand production management, where approaches such as laboratory physical simulation, numerical simulation, geomechanics, fluid mechanics, and computational mathematics (finite element and discrete element) are applied comprehensively, including the trend of stress around wellbore changes, critical condition of sanding, influence of sand production on the seepage conditions, sand production rate prediction at different stages, and so on.

6.1.2 Well Completion and Artificial Lift Techniques

The application of sand production management in heavy oil reservoirs must consider the maximum sand-carrying rate and the diameter of sand particles. Based on the requirements of production rate, it will provide the data of sand particle size for sand exclusion to analyze the sand-carrying capacity by conducting laboratory simulation experiments on different producing conditions and different heavy oil samples. For artificial lift, PCP can transport fluids with high sand concentration, and the maximum sand concentration can be 40%. Reasonable depth of pumps should be considered when producing with sand. When the depth of pump is too shallow, sand particles will settle down and the producing intervals will be buried because the flowing velocity in casing is slower than that in tubing. If the depth of pump is too deep and the produced sands cannot be pumped out, it is possible that the pump will be buried. Therefore, the depth of pumps should be optimized on the basis of researches of sand-carrying capacity of wellbore flowing.

6.1.3 Surface Treatment Techniques

Sand production management must consider the surface acceptable maximum sand handling capacity. In order to prevent produced sand particles from clogging and eroding storage and transportation system when producing with sands, the produced fluid will not flow into the pipe network before desanding. Because heavy oil can suspend sands for a longer period, the method of desanding using heating large tanks is applied to accelerate the settling velocity of produced sands. For instance, Suncor configured a large tank of 120 m^3 for each producing well at the beginning, and installed a 152-mm U-shaped heating pipe in the tank. The produced fluid flows into the tank, then to the pipe network after settling produced sands by heating. Later, in order to meet the increase of production rate, three big tanks were configured for each well, wherein two 152-mm heating pipes were installed in the first tank, and one heating pipe was installed in the second and the third tanks, respectively. The sand concentration reached the standard of purifying treatment and the produced fluid flows directly to the pipe network of external transportation after the three stages of treatment.

6.1.4 Comprehensive Economical Evaluation

The evaluation includes estimations of well completion types and costs, surface treatment costs, and production rates. The comprehensive economic benefits can be achieved on the basis of different contents of research and different calculating methods of costs during the thorough appraisal.

6.2 Implementation Workflow of Sand Production Management

Sand production management is a complicated and systematic technique throughout the development of reservoirs, and the workflow of execution is shown as Figure 6-1.

6.2.1 Reservoir Simulation Appraisal

Reservoir simulation appraisal includes the calculation of the distribution of pressure of pores and fluid saturation by conducting full oilfield simulation, appraising the trend of effective stress changes. With the help of rock mechanics and fluid mechanics calculation, it will provide the basis for the appraisal of the changes of formation flowing conditions, the status of sand production, the prediction of the volume of sand production, sand-carrying capacity in wellbore, and the optimization of well completion.

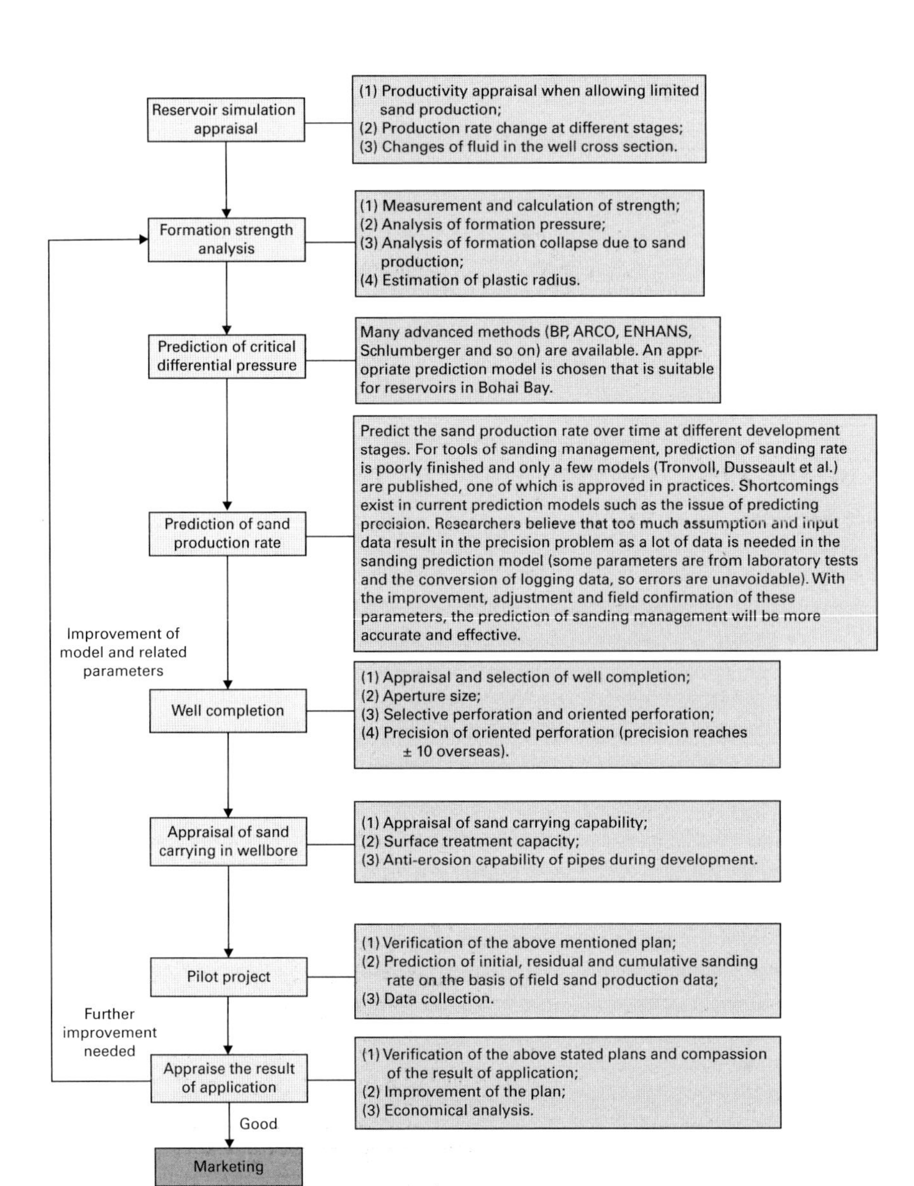

Figure 6-1 Sand production management patterns in Bohai oilfields.

6.2.2 Analysis of Formation Strength

Producing with sand will lead to cavities in formation and result in the changes of rock strength during development. Under the worst situations, collapse and subsidence of the formation will occur and the improvement of productivity index cannot be achieved. Therefore, scientific demonstrations should be provided and the following studies should be conducted before applying sanding management:

1. Analyze the mechanism of sand production in formation, and measure critical status of sand production of rock.
2. Measure rock strength, and appraise cohesion and internal friction angle of rocks by combining failure criteria.
3. Calibrate rock strength profile calculated by well logging data on the basis of experimental results of cores strength and obtain the data of elastic modulus, Young's modulus, and Poisson ratio.
4. Calculate the vertical stress of formation using density logging data, and obtain the horizontal stress by testing the breakdown pressure.

6.2.3 Selection of Well Completion

A suitable well completion type should be chosen on the basis of systematically analyzing reservoir characteristics, the distribution of movable sand particle sizes, the stage of reservoir development, sand-carrying capacity of wellbore, approaches of artificial lift, and the condition of surface treatment.

6.2.4 Quantitative Evaluation of the Risk of Sand Production

First, identify the critical status of sand production on the basis of the researches of rock mechanics. Then, calculate the profile of critical status of sand production according to the well logging data, and calibrate the profile using the core strength experiments. Analyze the risk of sanding under different pressure control and the potential sanding formation by integrating the critical condition of sand production, producing interval of wells, artificial lift, and operating parameters of wells.

6.2.5 Prediction of Initial Condition of Sand Production

Establish dynamic models of predicting the extension of plastic zone for vertical wells, deviated wells, and horizontal wells respectively on the basis of the distribution of effective stress surrounding the wellbore. Considering the sand production caused by the shear failure, the critical condition of sand production in different formations for given wellbore and reservoir pressure is established, and the continuous curve of critical condition of sand production with the changes of depth and bottom hole pressure is given.

6.2.6 Prediction of the Volume of Sand Production

Predict the relationship between the volume of sand production and pressure drop/time when the drawdown is greater than the critical pressure differential of sand production. The changes of volume of sand production with different producing strategy and well completion are the key factors of appraising the feasibility of sand production management.

6.2.7 Estimation of Sand-Carrying Capacity of Wellbore

In order to reduce the risk of the clogging by massive catastrophe-produced sands in the perforated tunnels and wellbore, you must make sure that the produced sand particles will be lifted to the surface instead of settling in the perforated tunnels and wellbore.

For vertical wells and slightly deviated wells, on the basis of well productivity, tubing size, well type, well completion type, developing approaches, sand particle size, sand concentration, risk estimation of producing with sands, etc., the limit of pressure differential is determined for given fluid components using established model based on Stokes criteria, and the integrated research of sand carrying in wellbore is conducted. For horizontal wells and highly deviated wells, the interaction among sand particles makes the model of sand particle migration complicated, and the quantitative description of it is still at the stage of theoretical research at present.

At the same time, the empirical criteria achieved from laboratory experiments is preferred in the petroleum industry, mainly used in designing the minimum flowing rate to prevent produced sands from building stable sand beds.

In addition, the monitoring plan of the surface of sanding formations should be designed.

6.2.8 Evaluation of the Erosion Rate of Devices

The purpose of risk analysis of erosion ratio is to determine the maximum production rate under which excessive erosion occurs. The major factors that are affecting the erosion of devices are flowing velocity, fluid density, sand particle sizes, volume of sand production, the diameter of transportation pipe and the stiffness of pipelines. Studies indicate that the erosion processes are with the following features:

1. The smaller the diameter of pipeline, the higher is the erosion ratio.
2. The higher concentration of gas, the higher is the erosion ratio.
3. The erosion ratio will increase with the increase of flowing rate when the flowing velocity is low, which is resulting from the increasing collision of sand particles on the lower part of pipelines where sand particles change from depositional status to flowing bed status.
4. With the increase of flowing velocity, sand particles will suspend and their collision with pipelines reduces, so the erosion ratio decreases.
5. When the flow velocity increases further, the erosion ratio will increase again because the energy of moving sand particles increases.

6.2.9 Surface Treatment Techniques when Producing with Limited Sand Production

The application of sand production management has to consider the maximum allowable capacity of surface treatment of produced sands when environment permits. When using the technique of producing with allowable sands, the produced fluid will contain some sand particles which can possibly block the gathering and transporting system and, therefore, the produced liquid cannot be flowed into the pipeline network before desanding on site.

The produced sand particles flow through the wellhead first, then through surface pipelines (through subsea pipelines for offshore oilfields) to the separator for precipitation. The separators should be cleaned and flushed regularly according to the estimated average volume of sand production. The produced sands containing oil should be collected for final treatment.

6.2.10 Pilot Application of Sand Production Management

The pilot testing of sand production management provides further understanding on the sanding features of formation. For instance, the field data of sand production in North Sea oilfields show the approximately parabolic relationship of the volume of sand production rate over time. Therefore, the obtained data from pilot wells are valuable for the study and design of sand production management, and the pilot test in field is a key step before applying this technique widely in all oilfields. Also, it must be pointed out that the pilot wells may fail due to improper design, resulting in the settling of sand particles in wellbore and expensive workover cost on offshore platform, even the risk of shutting in wells or production time loss. Therefore, careful selection of wells, formations and designs for pilot tests are very important.

6.3 Application of Sand Production Management in Oilfields of Bohai Bay

6.3.1 Introduction of Oilfields

The work area applying sand production management is the second zone of Group Dongying, Paleogene in Luda 5-2 oilfield. The depth of the oil pay is 1300 to 1700 m, the oil viscosity at surface is 760 to 8161 mPa·s, and the oil viscosity at subsurface is 210 to 400 mPa·s. The oilfield is with high gas–oil ratio and strong driving energy.

On the basis of the measurement of rock strength, the UCS (uniaxial compressive strength) is 3 to 5 MPs, and the formation is poorly consolidated and with low rock strength, belonging to potential sanding formation (Table 6-1).

6.3.2 Model of Sand Production Management

On the basis of studies of reservoir appraisal, the reservoir is with high gas oil ratio, strong driving energy and pretty good properties such as permeability and porosity. The content of clay is between 5% and 35%. The range of porosity is between 15% and 35% (mainly between 25% and 35%), and the average value of porosity is 30%. The variation of permeability is between 7,000 and 8,000 mD. The range of oil density is between 0.852 and 1.02 g/cm^3 (typically between 0.95 and 0.98 g/cm^3).This reservoir belongs to a conventional heavy oil reservoir.

According to the standards of appraising reservoirs proposed by J. Tronvoll and M.B. Dusseault, block LD 5-2 is the most suitable reservoir for applying the technique of sand production management. The rock strength is weak and even poor, and therefore, producing with limited sands will improve well productivity (Table 6-2).

As such, it is believed that it will improve well productivity of this reservoir to produce by allowing limited sand production.

Due to poor rock strength in the second zone of Group Dongying of LD 5-2 oilfield, sand production is unavoidable. The conventional sand exclusion using gravel packing is expensive and hard to operate with high skin and low productivity. The technique of CHOPS is also risky when implementing in

Table 6-1 UCS Prediction Result of Well LD 5-2-1

Formation	*Interval Depth (m)*	*Interval (m)*	*Interpretation Result*	*UCS (MPa)*	*Critical Differential Pressure (MPa)*
$E_3d_2^L$	1474.8 to 1476.8	2	Oil zone	2.34 to 2.67	1.17 to 1.33
$E_3d_2^L$	1594.4 to 1595.5	1.1	Oil zone	2.54 to 3.69	1.27 to 1.84
$E_3d_2^L$	1607.2 to 1612.3	5.1	Oil zone	2.02 to 5.02	1.01 to 2.51
$E_3d_2^L$	1612.8 to 1615.2	2.4	Oil zone	2.01 to 2.42	1.00 to 1.21
$E_3d_2^L$	1615.6 to 1618.4	2.8	Oil zone	2.11 to 2.71	1.05 to 1.35
$E_3d_2^L$	1620.7 to 1622.1	1.4	Oil zone	2.42 to 3.46	1.21 to 1.73
$E_3d_2^L$	1623.7 to 1625.5	1.8	Oil zone	2.64 to 3.25	1.32 to 1.62
$E_3d_2^L$	1626.9 to 1628.3	1.4	Oil zone	2.53 to 2.70	1.26 to 1.35
$E_3d_2^L$	1680.8 to 1682.6	1.8	Oil zone	3.71 to 9.69	1.85 to 4.85
$E_3d_2^L$	1686.3 to 1690.5	4.2	Oil zone	2.22 to 2.90	1.11 to 1.45

Table 6-2 Development Appraisal of Sand Production Management in LD 5-2 Oilfield

Content	*Data for Appraisal*	*Reservoir Conditions*
Reservoir type	Heavy oil reservoir	Conventional HO reservoir
Rock strength	2 to 10 MPa	3 to 5 MPa
Clay content	≤20% (preferably ≤10%)	Clay content 5% to 35%
Porosity	High porosity	25% to 35% (30% in average)
Permeability	High permeability	7000 to 8000 mD
Formation depth	300 to 1300 m	1300 to 1700 m
Formation pressure	—	11 to 13 MPa
GOR	High solution gas	GOR (m3/m3) 24.0
Thickness	≥3 m (preferably ≥5 m)	1.1 to 5.1 m
Viscosity	600 to 16,000 mPa·s (after degassing)	760 to 8161 mPa·s
Density	0.934 to 1.007 g/cm^3 (after degassing)	0.922 g/cm^3

offshore oilfield, resulting in massive sand production and difficult treatment of produced sands. Therefore, producing with allowable sands is beneficial for this kind of unconsolidated heavy oil reservoir.

On the basis of worldwide experiences and the actual data of oilfields in Bohai Bay, to encourage sand production using sand excluding screens is applied, where the aperture size of screen will be enlarged appropriately or drawdown will be enlarged. The principles of designing are as follows:

1. For well completion in oilfields in Bohai Bay, the technique of sand exclusion using screens is mature based on related experiences and similar and successful products.
2. To control sand with screens is commonly used in worldwide fields, such as Girassol oilfield and Darat Asset oilfield. The workshop of sanding management in offshore oilfields held in Houston in July 2004 proposed that one of the methods of sanding management is to extend the application scope of slotted liner and wire-wrapped screen.
3. The workspace is with low content of clay, suitable for completing with sand exclusion screens.
4. Compared with other techniques such as gravel packing, the application of screens is cheap, with high successful rate and low risk.
5. When conducting well completion with sand exclusion screens, the aperture has the function of self-cleaning because it is narrow inside and wide outside. Therefore, if the aperture is slightly blocked it can be fixed by the produced fluid.

6.3.3 Prediction of Volume of Sand Production

The predicted critical drawdown of LD 5-2 oilfield is 0.83 to 1.80 MPa, but the practical well drawdown is 2 MPa. Sand production will occur in most intervals of wells with this drawdown if sand exclusion is not applied. The prediction of the volume of sand production is conducted for open hole completion without sand control.

Based on the features of rock strength of the second zone of Dongying Group in LD 5-2 oilfield, the parameters chosen for the prediction of sand production are as follows:

The horizontal stress is 21 MPa, drainage radius is 175 m, initial reservoir pressure is 12.47 MPa, bottom hole flowing pressure is 10.47 MPa, radius of wellbore is 0.108 m, rock cohesion strength is 1 MPa, internal friction angle is 34°, Poisson's ratio is 0.3, elastic modulus is 3600 MPa, oil viscosity is 300 cP, initial permeability of formation is 7000 mD, and initial porosity is 30%.

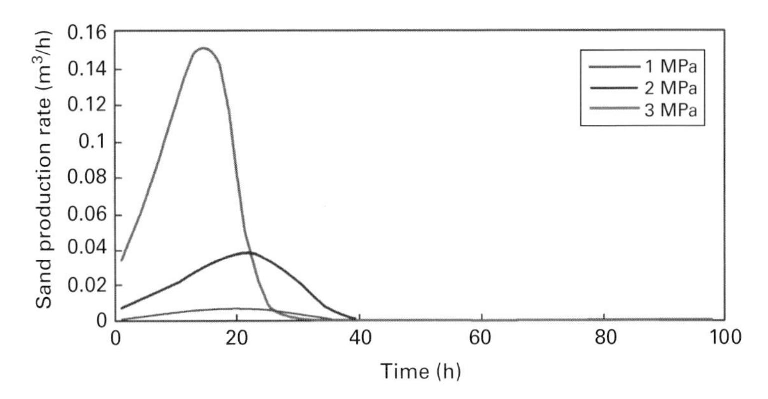

Figure 6-2 Volume of sand produced per unit thickness over time at different drawdown.

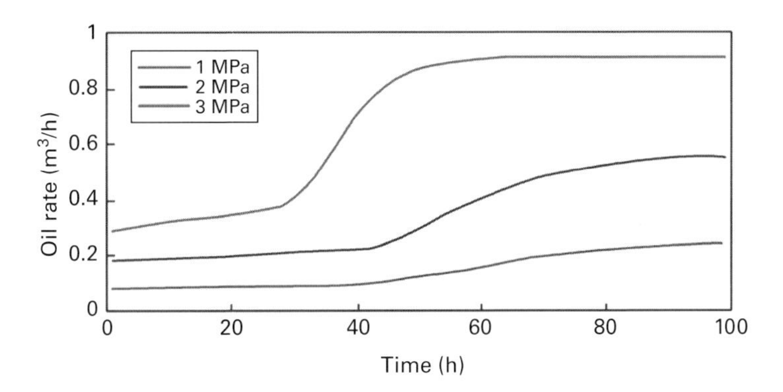

Figure 6-3 Oil production rate per unit thickness over time at different drawdown.

Based on above listed parameters, the predicted volume of sand produced is shown in Figure 6-2 and Figure 6-3. As it can be seen from the figures, the volume of sand produced is large at the beginning of production, and the maximum sand concentration can be greater than 19% with drawdown of 2 MPa. If keeping the operating parameters constant, the volume of sand produced will decrease after producing for a period of time and tend to be stable eventually. The result is consistent with the principles of sand production from laboratory experiments and practices of CHOPS. According to data of development of heavy oil fields worldwide, the sand concentration in produced liquid at the early stage of production applying CHOPS will be 20% to 60% by volume, and it will decrease to 0.5% to 3.0% and tend to be stable after 6 months to one year. But with the increase of production rate, sand concentration will possibly increase again within a short term, and the sand concentration will decrease to the previous level if keeping production rate constant.

6.3.4 Design of Aperture Sizes

The distribution of sand particle sizes of formation rock varies in a large range on the basis of analyzing the data of particle size of 18 samples from well LD 5-2-1 (d_{50} is 185 to 500 μm), so formation sands are inhomogeneous (the heterogeneous coefficient of sands is 1.88 to 6.54).

The design of aperture sizes of sand exclusion screens is conducted for well LD 5-2-1 using available designing methods of aperture size, and the results are shown in Table 6-3. It can be seen from the table that the difference of aperture sizes is large using different methods.

After comparing and analyzing different approaches, the Reslink software is used to calculate the relationship between cumulative mass percentage and particle sizes of 18 samples from Luda oilfield, and the Corberly method is used to design the aperture size (the aperture size is 2.5 times of the maximum particle size). The aperture size, sand concentration and the maximum particle size in tubing under different conditions of sand control are achieved (Figure 6-4 and Table 6-4).

According to the predicted volume of sand produced with 2 MPa drawdown, the sand concentration in sand exclusion screen is 1.94%, 2.91%, 3.88% and

Table 6-3 Aperture Sizes of Sand Exclusion Screens for Well LD 5-2-1 Using Different Approaches

Design Approach	*Descriptions*	*Aperture Size of Screen*
d_{10}	Used in onshore oilfield (e.g. Gulf of Mexico, US)	510 μm
$2d_{10}$	Used in onshore oilfield (e.g. California, US)	1020 μm
Con-slot	—	230 μm
Abrams	Settling of particles on the surface of screen (1/3 blocking and bridging principle)	315 μm (10%)
Corberly and Gillespie	2.5 times the maximum particle size	262.5 μm (10%)
Reslink	Analysis of major components	250 to 300 μm

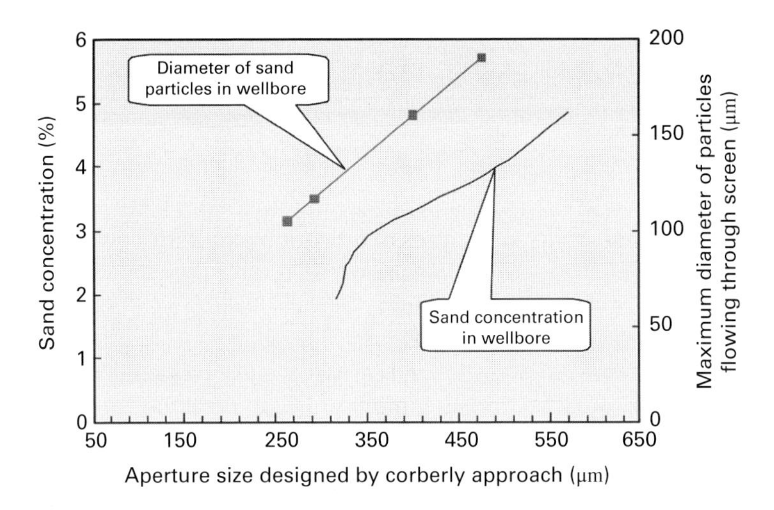

Figure 6-4 Required aperture size and the maximum sand particle size with different sand concentration in tubing.

Table 6-4 Required Aperture Size of Different Sand Concentration in Tubing

Cumulative Weight Percentile of Sands Excluded by Screen (%)	*Sand Concentration in Wellbore (%)*	*Maximum Diameter of Sand Particles (μm)*	*Aperture Size Designed by Corberly and Gillespie Approach (μm)*	*Aperture Size Designed by Abrams Approach (μm)*
90	1.94	105	262.5	315
85	2.91	117	292.5	351
80	3.88	160	400	480
75	4.85	190	475	570

4.85% when screen control formation sands by 90%, 85%, 80%, and 75%, respectively. The maximum sand particle sizes in tubing are 105, 117, 160, and 190 μm, as shown in Table 6-4.

6.3.5 Calculation of Sand Carrying in Wellbore

On the basis of the aperture size of sanding management applied in LD 5-2 oilfield, the calculation of critical sand-carrying parameters is conducted for the sand particles in the range of 105 to 190 μm. The oil viscosity is

210 mPa·s, tubing size is 73 and 88.9 mm, and casing size is 177.8 mm (Tables 6-5 to 6-7, Figures 6-5 to 6-7).

Calculation Results

The calculation results of sand-carrying capacity in wellbore based on the above parameters show that the key factors affecting critical sand-carrying production rate under certain conditions are the size of production string, well deviation and oil viscosity. The critical sand-carrying production rate of casing is 8 times that of tubing, critical sand-carrying production rate for wells with inclination 65° is 4 times that of vertical wells, and critical sand-carrying production rate of oil viscosity 15 mPa·s is 12 times that of 210 mPa·s. Friction loss due to the change of oil viscosity is less under critical sand-carrying production rate, but the pressure drop gradient resulting from particles concentration is larger. Taking the vertical wells as example, the pressure gradient increases significantly when the diameter of particles is larger than 400 μm. When considering the factors that will influence the well performance, the following conclusion was achieved that different operating conditions should be given for wells with various inclination to obtain the critical sand-carrying production rate, and the diameter of sands allowed to wellbore should

Table 6-5 Critical Sand-Carrying Production Rates with 73-mm Tubing

Cumulative Weight Percentile of Sands Excluded by Screen (%)	*Sand Concentration in Wellbore (%)*	*Maximum Diameter of Sand Particles (μm)*	*Inclination (°)*	*Production Rate for Sand Carrying (m³/d)*		
				15 mPa·s	*210 mPa·s*	*400 mPa·s*
90	1.94	105	0	0.678	0.048	0.025
			30	1.291	0.092	0.048
			65	1.426	0.102	0.053
			90	1.034	0.074	0.039
85	2.91	117	0	0.774	0.055	0.029
			30	1.507	0.108	0.056
			65	1.665	0.119	0.062
			90	1.164	0.083	0.044
80	3.88	160	0	1.212	0.087	0.045
			30	2.543	0.182	0.095
			65	2.809	0.201	0.105
			90	1.736	0.124	0.065
75	4.85	190	0	1.492	0.107	0.056
			30	3.290	0.235	0.123
			65	3.635	0.260	0.136
			90	2.072	0.148	0.078

Table 6-6 Critical Sand-Carrying Production Rate with 88.9-mm Tubing

Cumulative Weight Percentile of Sands Excluded by Screen (%)	*Sand Concentration in Wellbore (%)*	*Maximum Diameter of Sand Particles (μm)*	*Inclination (°)*	*Production Rate for Sand Carrying (m^3/d)*		
				15 mPa·s	*210 mPa·s*	*400 mPa·s*
90	1.94	105	0	0.942	0.067	0.035
			30	1.793	0.128	0.067
			65	1.981	0.141	0.074
			90	1.435	0.103	0.054
85	2.91	117	0	1.075	0.077	0.040
			30	2.092	0.149	0.078
			65	2.312	0.165	0.087
			90	1.616	0.115	0.061
80	3.88	160	0	1.685	0.120	0.063
			30	3.534	0.252	0.133
			65	3.905	0.279	0.146
			90	2.413	0.172	0.090
75	4.85	190	0	2.076	0.148	0.078
			30	4.575	0.327	0.172
			65	5.055	0.361	0.190
			90	2.882	0.206	0.108

Table 6-7 Critical Sand-Carrying Production Rate with 177.8-mm Casing

Cumulative Weight Percentile of Sands Excluded by Screen (%)	*Sand Concentration in Wellbore (%)*	*Maximum Diameter of Sand Particles (μm)*	*Inclination (°)*	*Production Rate for Sand Carrying (m^3/d)*		
				15 mPa·s	*210 mPa·s*	*400 mPa·s*
90	1.94	105	0	4.512	0.322	0.169
			30	8.589	0.614	0.322
			65	9.490	0.678	0.356
			90	6.878	0.491	0.258
85	2.91	117	0	5.157	0.368	0.193
			30	10.033	0.717	0.376
			65	11.085	0.792	0.416
			90	7.750	0.554	0.291
80	3.88	160	0	8.099	0.578	0.304
			30	16.985	1.213	0.637
			65	18.767	1.340	0.704
			90	11.597	0.828	0.435
75	4.85	190	0	9.993	0.714	0.375
			30	22.026	1.573	0.826
			65	24.336	1.738	0.913
			90	13.875	0.991	0.520

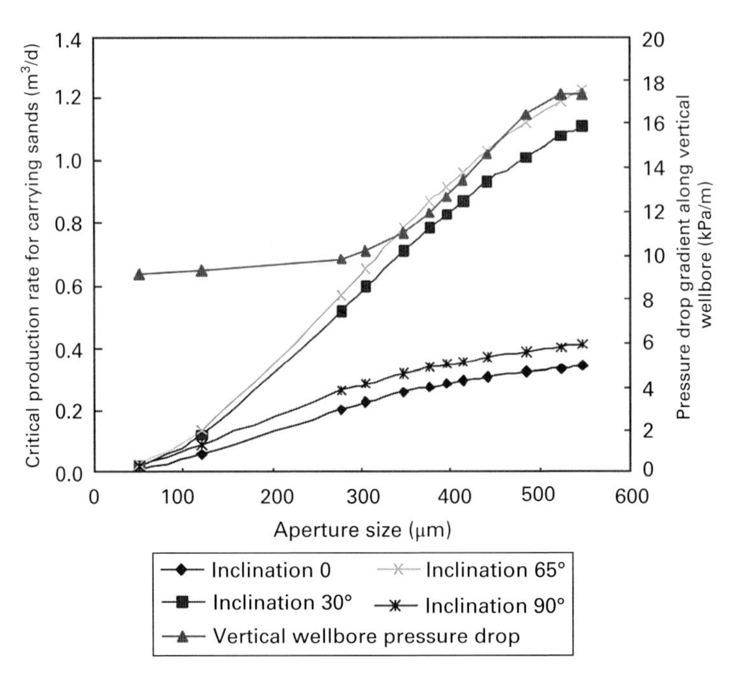

Figure 6-5 Critical production rate for safe sand carrying in 73-mm tubing with 210 mPa·s oil.

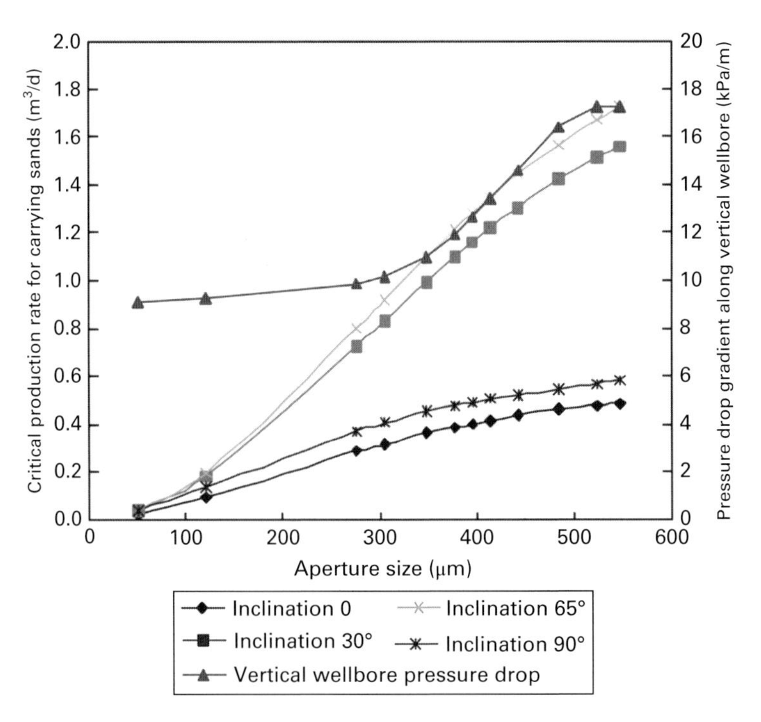

Figure 6-6 Critical production rate for safe sand carrying in 88.9-mm tubing with 210 mPa·s oil.

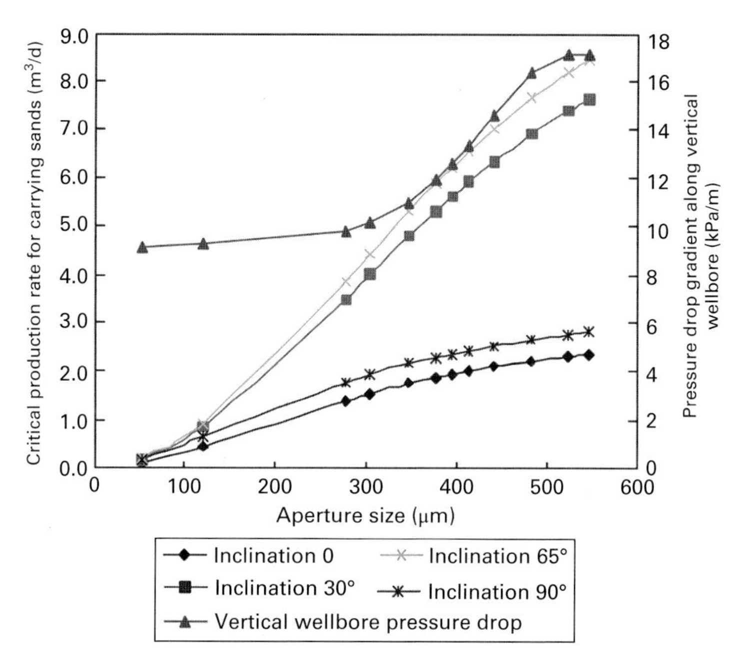

Figure 6-7 Critical production rate for safe sand carrying in 177.8-mm tubing with 210 mPa·s oil.

be controlled within 400 μm after integrating the factors such as pump efficiency and energy consumption.

6.3.6 Artificial Lift and Surface Oily Sands Treatment

6.3.6.1 Artificial lift

When producing with controlled sands in the LD 5-2 oilfield, the maximum sand concentration can be controlled under 5% w/w on the basis of the design of sand exclusion. PCP is very suitable for the reservoir with viscosity larger than 300 mPa·s, meeting the requirements of production rate.

The recommended plan of artificial lift method is to use submersible PCP for high angle inclination wells and to apply the surface driving PCP for the wells with inclination less than 40°.

6.3.6.2 Technique of encouraging sand production

After producing with designed aperture size for a period of time, the colloid and asphalt will deposit and some small-sized particles will settle around well-bore to form sand bridges, which results in the decrease of well productivity and production rate. Appropriate measures should be undertaken to encourage sand production to break the sand bridge and thus to improve well productivity and production rate. In LD 5-2 Oilfield, the existing pumps for killing wells and drilling pumps can be used to inject hot diesel or other chemicals to encourage sand production.

6.3.6.3 Treatment and discharge of surface oily sands

On the basis of oil properties of LD 5-2 oilfield, the desanding process of oil and the cleaning process of oil-contained sands are proposed in Figure 5-22. In the early stage of production, water from source well, viscosity reducer, emulsify and so on are added to flowing process to maintain daily production when there is no or small amount of water rate. When the water rate is high enough, the water source well can be closed. The produced sands can be discharged to sea directly after treating and meeting the standards, and it is not necessary to transport the produced sands to land for treatment.

Bibliography

BAOHE, W. & ZHONGXI, W. (1996) A new method to calculate the free settling velocity of spherical particles. *Power Technology*. 2 (2).

BEHRMANN, L. A., WILLSON, S. M., BREE, P. H. & PRESLES, C. (1997) Field Implications for Full-Scale Production Experiments. SPE 38639 presented at the 1997 SPE Annual Technical Conference and Exhibition in San Antonio, October 5–8.

BOYING, S. (1989) The settling sands and desanding of heavy and viscous oil overseas. *Huabei Petroleum Design*. 5 (1) p.32–35.

BROWN, J. E., KING, L. R. & NELSON, E. B. (1996) Use of a Viscoelastic Carrier Fluid in Frac-Pack Applications. SPE 31114 presented at the 1996 SPE Formation Damage Symposium in Lafayette, LA, February 14–15.

CHENGJIAN, L. (2004) Experimental studies of the influence of crude viscosity containing water on the performance of ESP. *China Offshore Oil and Gas*. 16 (2).

CHENGJIAN, L. (2007) The proposal of desanding of oil with sands and cleaning sands with oil techniques of offshore heavy oil fields. *China Offshore Oil and Gas*. 19 (4).

CHENGJIAN, L., et al. (2001) The application of one VSD driving multi-ESPs technology (OVDMT) in offshore heavy oil field. *SPE* 64619.

CINCOLEY, H., SAMANIEGO, V. & DOMINGUEZ, N. (1978) Transient pressure behaviour for a well with a finite conductivity vertical fracture. *SPEJ*. p.253–264.

DALIN, L., JUN, Z. & YINGMING, F. (2002) The application of high-pressure desanding devices in surface testing of oil and gas wells. *Well Testing and Production Technology*. 20 (3) p.56–58.

DONGLIN, L. & YUQI, P. (1997) A new type degritting technique for closed container. *Oil-Gasfield Surface Engineering*. 16 (1) p.22–24.

EDITORIAL COMMITTEE OF *CHEMISTRY ENGINEERING HANDBOOK*. (1989) *Chemistry Engineering Handbook*. Beijing: Chemical Industry Press.

Sand Production Management for Unconsolidated Sandstone Reservoirs, First Edition.
Shouwei Zhou and Fujie Sun.

FENGJI, Y., MINGYI, L. & YINGLAI, S. (1995) Integrated treatment technique of producing-gathering-transporting-refining of heavy oil field. *Oil-Gasfield Surface Engineering.* 14 (2) p.5–10.

GEILIKMAN, M. B., DUSSEAULT, M. B. & Dullien, F. A. L. (1998) Dynamics of Wormholes and Enhancement of Fluid Production, Tech. Rep., Part I, Waterloo Sand Production Project Report.

GENGSHENG, H. (1994) *Petrophysics.* Beijing: Petroleum Industry Press, 179–.

GUANGWEN, C. & DESHENG, G. (1994) Exploration of drag losses calculation in slurry horizontal pipeline transportation. *Journal of Central South University.* 25 (2).

HONG, T., JINGENG, D., BAOHUA, W., et al. (2006) Sand production prediction technology review of weakly consolidated sand reservoir. *Petroleum Geology & Oilfield Development in Daqing.* 25 (2) p.61–64.

HONGLIE, Z. (1995) Study on transporting parameters of coal slurry pipes using hydraulic coal mining. *Hydraulic Coal Mining & Pipeline Transportation.* 4.

HONGXIANG, G., CHENGJIAN, L., et al. (2008) Research of production rate adjustment capability of ESP wells under the condition of variable frequency. *Fluid Machinery.* 36 (2).

JIANPING, S. (2005) *Studies of the Mechanisms of CHOPS in Unconsolidated Sandstone Heavy Oil Reservoirs.* Chengdu: Southwest Petroleum University.

JIANXIN, X., JINREN, N. & JIAZHEN, H. (2002) Pressure loss in solid-liquid flow with coarse manganese nodules in vertical pipeline. *Journal of Sediment Research.* (2).

MAY, E. A., BRITT, L. K. & NOLTE, K. G. (1997) The Effect of Yield Stress on Fracture Fluid Clean-up. SPE 38619 presented at the 1997 SPE Annual Technical Conference and Exhibition in San Antonio, Texas, October 5–9.

NING, W., QI, Z. & ZHANGQING, Q. (2000) Evaluation on calculation methods of solid particle settling velocity in fluid. *Oil Drilling & Production Technology.* 22 (2).

PAPAMICHOS, E. & MALMANGER, E. M. (2001) A Sand-Erosion Model for Volumetric Sand Predictions in a North Sea Reservoir. SPE 69841-PA, p.44–50, February.

QIQUAN, R. & SHILUN, L. (1997) Study on dynamic models of reservoir parameters when performing coupled simulation. *Petroleum Exploration and Development.* 24 (3) p.61–65.

SHANGHAI CHEMICAL INSTITUTE, CHENGDU UNIVERSITY OF SCIENCE AND TECHNOLOGY, DALIAN UNIVERSITY OF TECHNOLOGY. (1980) *Chemical Engineering*. Beijing: Chemical Industry Press, p.67–106, 213.

SHAOZHI, C., CHANGZHONG, H. & XINFU, L. (1997) *Technologies of Cold Heavy Oil Production with Sand.* Beijing: Petroleum Industrial Press, p.3–39.

SHOUWEI, Z. (2002) *Study on Strategy and Management of Developing Offshore Oil Resources in China*. Chengdu: Southwest Petroleum University.

SHOUWEI, Z. (2007) *Probe and Application of the New Model to Develop Offshore Oilfields Efficiently.* Beijing: Petroleum Industry Press, p.126–128.

TREMBLAY, B., SEDGWICK, G. & FORSHNER, K. (1997) Simulation of Cold Production in Heavy-Oil Reservoirs: Wormhole Dynamics. SPE 35387- PA, 110–117, May.

WANG, J. & WAN, R. G. (2004) Computation of sand fluidization phenomena using stabilized finite elements. *Finite Elements in Analysis and Design.* 40 p.1681–1699.

WENBIN, J., XIAOPING, H. & WENPENG, G. (1997) An experimental study on minimal transporting velocity of the lifted large-size particle material in vertical piping. *Mining Research and Development.* 17 (2).

WENXING, W. & XIUPING, C. (2001) Design of sludge discharging system of super viscous oil & sewage deoiling tank. *Oil-Gasfield Surface Engineering.* 20 (2) p.34.

WENXING, W. & XIUPING, C. (2001) The study of dewatering techniques of high viscosity crude. *Oil-Gasfield Surface Engineering.* 16 (1) p.19–21.

XIANGJUN, F. (2000) Physical characteristics and flowing velocity transferred by pipeline of slurries. *Pipeline Technique and Equipment.*

XIANGLIN, Z., FUJIE, S., GUANJUN, H., et al. (2004) To improve well productivity by allowing controlled sands in SZ36-1 heavy oil reservoir. *Special Oil & Gas Reservoirs.* 11 (6) p.47–49.

XIANGLIN, Z., FUJIE, S., XING, W. & PINGSHUANG, W. (2005) Well productivity research on conventional heavy oil in unconsolidated reservoirs in Bohai Oilfield. *Oil Drilling & Production Technology.* 28 (6) p.36–39.

XIANGLIN, Z., GUANJUN, H., FUJIE, S., et al. (2005) Influences of sand production on permeability and experiments on sand production characters in SZ36-1 Oilfield. *Petroleum Exploration and Development.* 32 (6) p.105–108.

YANHUA, S., JILING, M. & SHENG, L. (1999) Minimum lifting water velocity of solid-liquid two-phase flow in vertical pipe. *Journal of University of Science and Technology Beijing.* 21 (6).

YANHUA, S., JILING, M. & SHENG, L. (2000) The simulation of artificial lifting parameters of inclined lifting minerals pipes in offshore mining. *Nonferrous Metals (Mining Section).* (1).

YARLONG, W. (2001) A coupled reservoir-geomechanics model and applications to wellbore stability and sand prediction. *SPE* 69718. March.

YARLONG, W. (2001) An integrated reservoir model for sand production and foamy oil flow during cold heavy oil production. *SPE* 69714. March.

YARLONG, W. (2002) Coupled reservoir-geomechanics model with sand erosion for sand rate and enhanced production prediction. *SPE* 73738. February.

YONGLE, H., QIQUAN, R. & JIANPING, S. (2007) *Fluid-Solid Coupling Numerical Simulation.* Beijing: Petroleum Industry Press.

ZEMIN, Z. (2002) Exploring to heavy crude dehydration. *Oil-Gasfield Surface Engineering.* 19 (2) p.31–32.

ZHENGMAO, W. (2004) *The Study of Fluid Flow Mechanism with Sand Erosion and Sand Particulates Migration in the Reservoir and Fluid-Solid Coupling Single-Well Numerical Simulation.* Chengdu: Southwest Petroleum University.

Index

Sand Production Management for Unconsolidated Sandstone Reservoirs, First Edition.
Shouwei Zhou and Fujie Sun.